Scattering Theory

Pure and Applied Mathematics

A Series of Monographs and Textbooks

Edited by

Paul A. Smith and Samuel Eilenberg

Columbia University, New York

Pure and Applied Mathematics

A Series of Monographs and Textbooks

In memory of
Franz Rellich

SCATTERING THEORY

PETER D. LAX

COURANT INSTITUTE OF MATHEMATICAL SCIENCES
NEW YORK UNIVERSITY
NEW YORK, NEW YORK

RALPH S. PHILLIPS

DEPARTMENT OF MATHEMATICS
STANFORD UNIVERSITY
STANFORD, CALIFORNIA

1967

ACADEMIC PRESS   New York and London

ACADEMIC PRESS INC.
111 Fifth Avenue, New York, New York 10003

United Kingdom Edition published by
ACADEMIC PRESS INC. (LONDON) LTD.
Berkeley Square House, London W.1

LIBRARY OF CONGRESS CATALOG CARD NUMBER: 66-30098

PRINTED IN THE UNITED STATES OF AMERICA

Preface

Scattering theory compares the asymptotic behavior of an evolving system as t tends to $-\infty$ with its asymptotic behavior as t tends to $+\infty$. It is especially fruitful for studying systems constructed from a simpler system by the imposition of a disturbance (also called perturbation or scatterer) provided that the influence of the disturbance on motions at large $|t|$ is negligible, i.e., if any motion of the *perturbed system* for large $|t|$ is indistinguishable from a motion of the *unperturbed system*. Thus, if $U(t)$ and $U_0(t)$ denote the operators relating the states of the perturbed and unperturbed systems at time zero to their respective states at time t, then to each state f of the perturbed system there correspond two states f_- and f_+ of the unperturbed system such that $U(t)f$ behaves like $U_0(t)f_-$ as $t \to -\infty$ and like $U_0(t)f_+$ as $t \to +\infty$. The scattering operator is defined as the mapping:

$$S: f_- \to f_+ .$$

The aim of scattering theory is to prove the existence of such a scattering operator and to link its properties to the nature of the scatterer. In situations where the scattering operator constitutes the only physically observable data of motion the main task is the *inverse problem* of reconstructing the scatterer from the scattering operator.

This notion of scattering is meaningful for systems described by nonlinear operators. However, most work on scattering theory, including the present monograph, deals with linear time-invariant systems in which case $\{U(t)\}$ form a one-parameter group of linear operators.

In our approach we deal with systems described by a group of unitary operators $\{U(t)\}$ acting on a Hilbert space H in which there are two distinguished subspaces D_- and D_+, with the property that, as t varies from

$-\infty$ to $+\infty$, the subspaces $U(t)D_-$ and $U(t)D_+$ increase (decrease) mono-tonically from the zero subspace to the whole space H; we call D_- and D_+ the incoming and outgoing subspaces, respectively. It is not difficult to show that with each subspace D_- and D_+ we can associate a special spectral representation of the group $\{U(t)\}$; in the one D_- is represented by func-tions analytic in the lower half-plane, in the second D_+ is represented by functions analytic in the upper half-plane. The two representations are related by a unitary, operator-valued multiplicative factor $S(\sigma)$, $-\infty <$ $\sigma < \infty$, which we call the *scattering matrix*. If D_- and D_+ are orthogonal then $S(\sigma)$ is the restriction to the real axis of a bounded operator-valued analytic function homomorphic in the lower half-plane.

We apply this theory to systems governed by hyperbolic differential equations. The unit form of the Hilbert space is defined as energy; D_- con-sists of all initial states f such that $U(t)f$ is zero in some backward cone $|x| < -ct + \rho$, D_+ of states f for which $U(t)f$ is zero in some forward cone $|x| < ct + \rho$. Here ρ is so chosen that all scatterers, i.e., obstacles, potentials, and inhomogeneities, are contained in the ball $\{|x| < \rho\}$. We show that in an odd number of space dimensions D_+ and D_- are orthogonal.

Denote by P_+ and P_- the operators which remove the D_+ and D_- com-ponents, i.e., project onto the orthogonal complements of D_+ and D_-. Since incoming motions are not influenced by the scatterer for $t < 0$, and outgoing motions are not influenced for $t > 0$, they can be discarded without losing any information about the scatterer. This suggests looking at the operators

$$Z(t) = P_+ U(t) P_- \qquad (t > 0).$$

We show that these operators form a semigroup closely connected with the scattering matrix: the set of points in the lower half-plane at which the scattering matrix is not invertible is $i\Sigma$, where Σ denotes the spectrum of the infinitesimal generator of $\{Z(t)\}$.

We use the theory of hyperbolic equations to study directly the operators $\{Z(t)\}$; the information obtained this way relates the spectrum of the infinitesimal generator to geometric and analytic properties of the obstacle. We can show that different obstacles give rise to different scattering matrices. Using the Birman–Kato principle of the invariance of the scat-tering operator, we obtain the scattering matrix for the Schrödinger equation from the scattering matrix for the corresponding wave equation.

Our theory is applicable only to perturbations which act in a bounded domain and in spaces of odd dimension. In this respect our approach is more restricted than the time dependent scattering theory of Møller [1], Rosenblum [1], Kato and Kuroda [1]. On the other hand we have a means of studying the scattering operator directly.

The steady state theories of Friedrichs and of Lippman-Schwinger [1] offer another way of studying directly the scattering operator; using this approach Carleman, Povsner [1], Ladyzhenskaya [1], Mizohata [1] and recently Dolph, McLeod and Thoe [1] were able to deduce various analytic properties of the scattering matrix. Also in this way Faddeev [1] has succeeded in proving the existence of the scattering matrix for the three-body problem.

The main tools in the steady-state theories are the theory of elliptic equations and integral equations, whereas our approach is based on the theory of hyperbolic equations and the theory of invariant subspaces initiated by Beurling. We also make use of the Radon transform to give a new (and it seems to us more natural) formulation of the Sommerfeld radiation condition applicable also to general hyperbolic equations.

An interesting generalization of our theory has been given by Adamjan and Arov [1, 2].

We acknowledge with pleasure our debt to Cathleen Morawetz; a part of the results described in this monograph were obtained jointly with her. We also thank Professor M. G. Krein for his interest in our work.

We thank the AEC, NSF, ONR, and Sloan Foundation for their support and encouragement in carrying out this work.

We thank Isolde Field for her capable and devoted typing of our manuscript.

We dedicate this monograph to the memory of Franz Rellich, not only because his compactness criterion and uniqueness theorem are central to our work, but also as recognition of his pioneering role, along with Friedrichs, in using the methods of functional analysis and the theory of partial differential equations for attacking problems of mathematical physics.

P. D. Lax
R. S. Phillips

March, 1967

Contents

CHAPTER I

Introduction

In this monograph we shall investigate classical—in contrast to quantum mechanical—scattering theory. A prototype is the theory of scattering of sound waves in three-dimensional space by an obstacle. The propagation of sound waves is governed by the wave equation

$$u_{tt} - \Delta u = 0 \; ;$$

the amplitude u satisfies this equation outside the obstacle while on the obstacle u is required to be zero. The total energy contained in the solution at time t is

$$\frac{1}{2} \int \left[|\, \partial_x u \,|^2 + |\, \partial_t u \,|^2 \right] dx,$$

where the integration extends over the exterior of the obstacle. This quantity is independent of time. In what follows we shall deal mostly with solutions having finite energy.

It is intuitively plausible—although not at all easy to prove—that for any solution very little energy remains in regions near the obstacle as $|\, t \,|$ becomes large. This indicates that the obstacle has very little influence on the solution when $|\, t \,|$ is large. This is indeed the case, in the following specific sense:

Given any solution u of the wave equation outside the obstacle, vanishing on the obstacle and having finite energy, there exist two solutions u_- and u_+ of the free space wave equation—that is, with no obstacle present— such that, as t tends to $-\infty$ [or to $+\infty$], the energy of the difference $u - u_-$, respectively $u - u_+$, tends to zero. Thus the effect of the obstacle is to change a free space solution which starts out as u_- near $t = -\infty$ into the free space solution u_+ near $t = +\infty$. The operator S relating u_- to u_+ is called the *scattering operator*.

The importance of the scattering operator lies in this: if the obstacle is remote or is otherwise inaccessible to direct observation, then one of the principal methods for investigating its shape is to examine the transmitted and reflected waves of various frequencies. This is similar to the interaction of atomic nuclei with waves or particles where the time scale is such that observations made before and after an experiment can be considered as occurring near $t = -\infty$ and $t = +\infty$. As explained in Section 4 these transmitted and reflected waves are closely connected with the scattering operator. Thus in these situations the scattering operator is the basic observable of the interaction; and it is for this reason that Heisenberg, the founder of the modern quantum theory of scattering, conjectured that all the pertinent information about nuclear forces is contained implicitly in the scattering operator. One of the basic problems in this theory is that of extracting this information; this is the so-called *inverse scattering problem* and it has occupied the attention of many physicists and mathematicians. Our only contribution to this fascinating topic is a proof, presented at the end of Chapter V, that for the acoustic equation the scattering operator uniquely determines the shape of the scattering obstacle.

Some insight into the acoustic scattering problem can be gained from the transport problem for the same exterior domain; here the propagation of sound waves is replaced by a flow of particles along straight lines, except that when encountering the obstacle the particles bounce off according to the law of reflection: the incident and reflected rays lying in a plane containing the normal to the obstacle at the point of incidence and forming equal angles with this normal. In this corpuscular model, which we treat in Section 6, the nature of the scattering process depends strongly on the geometry of the obstacle. Thus, there is a sharp dichotomy in the scattering behavior according as the configuration does or does not satisfy the following property: The sojourn times of all trajectories within some ball containing the object is bounded. We conjecture, but do not quite prove, that this property has a similar influence on the scattering associated with the exterior problem for the wave equation. That is we expect, when the object satisfies this property, that the main interaction lasts only a finite length of time and eventually all signals decay exponentially at every point outside of the obstacle. The gap in our proof is technical, due to unresolved difficulties in constructing diffracted fields near the shadow boundary; but we are able to prove the result for star-shaped obstacles.

The rest of this chapter is devoted to a more detailed description of the contents of this monograph.

In Section 1 we give a rigorous definition of the scattering operator in terms of the wave operators introduced by Jauch and Møller. In Section 2 we present an alternative formulation of scattering theory in terms of a correspondence between two distinguished spectral representations of the one-parameter group of unitary operators determined by the solutions to the exterior problem. In Section 3 we relate scattering theory to one-parameter semigroups of contraction operators and also to the theory of translation invariant subspaces. This semigroup of contraction operators reveals the connection between the local asymptotic behavior of the solution and the exponentially increasing eigenfunctions associated with radioactive states in quantum mechanical scattering. In Section 4 we connect the above dynamic scattering theories to the more classical normal mode analysis and exhibit several equivalent formulations of the Sommerfeld radiation condition. In Section 5 we give a detailed analysis of a particularly simple one-dimensional example and in Section 6 we sketch the application of our theory to the transport problem in an exterior domain.

The theory sketched in the first chapter is developed at length in Chapters II to V. All of the really hard analysis is relegated to Chapter V; in particular our approach leads to a very clean proof of the existence of solutions of the reduced wave equation in an exterior domain with given boundary values and satisfying the radiation condition near infinity.

Chapter VI presents, somewhat sketchily, further applications of the abstract theory to symmetric hyperbolic systems and the acoustic equation with an indefinite energy form. We are then able to obtain the scattering operator for the Schrödinger equation by relating this problem to that of the acoustic equation with a potential.

The first appendix gives the necessary background in the theory of semigroups of operators; the second appendix contains two alternative derivations of the crucial energy decay theorem; the third appendix by C. S. Morawetz consists of a new derivation of certain identities and inequalities needed in the study of scattering by star-shaped obstacles; and the fourth appendix by G. Schmidt indicates how the electromagnetic equations can be treated within the framework of our theory.

1. The Dynamic Approach

In the dynamic theory of scattering the behavior for large times of solutions to the wave equation outside an obstacle is compared with the corresponding behavior of solutions in free space; the former is called the perturbed system and the latter the unperturbed system.

According to the classical theory of the wave equation a solution u in free space is uniquely determined by the values of u and u_t at any time $t = T$; we call the pair of functions $\{u(T), u_t(T)\}$ the *data* of u at time T. In particular the data at time zero are called initial data and are denoted as $f = \{f_1, f_2\}$, where $f_1 = u(0)$ and $f_2 = u_t(0)$.

The operator relating initial data to data at time t will be written as $U_0(t)$; these operators form a one-parameter group. Moreover, *energy* is conserved, that is the quantity

$$\frac{1}{2} \int \left[|\, \partial_x u(T) \,|^2 + |\, \partial_t u(T) \,|^2 \right] dx$$

is independent of T. Hence if we measure data by the energy norm:

$$(1.1) \qquad\qquad |f|_{E^2} = \frac{1}{2} \int \left[|\, \partial_x f_1 \,|^2 + |\, f_2 \,|^2 \right] dx,$$

the operators $\{U_0(t)\}$ are isometric, actually *unitary* since they are also invertible. The set of all data f with finite energy defines the Hilbert space H_0.

A similar description can be given of solutions of the wave equation outside the obstacle which satisfy a homogeneous boundary condition such as the requirement that u vanish on the obstacle. Again, data at any time T given outside the obstacle determine a solution uniquely. The operator relating initial data to data at time t will be written as $U(t)$; these operators also form a one-parameter group. Here too energy is conserved; however for more complicated boundary conditions the expression for total energy may include a boundary term. Disregarding this possibility we again introduce (1.1) as the energy norm, where now the integration extends only over the exterior of the obstacle. The operators $\{U(t)\}$ are unitary with respect to this energy norm on the Hilbert space H of all data defined outside the obstacle and having finite energy. It is convenient to regard data in H_0 which vanish near the obstacle as also belonging to H; this can be accomplished by merely restricting the support of such data to the exterior of the obstacle.

The principal tools for comparing the behavior of the perturbed and unperturbed solutions are the *wave operators* W_+ and W_- defined as follows:

$$(1.2) \qquad\qquad W_+ f = \lim_{t \to \infty} U(-t) U_0(t) f,$$

$$(1.3) \qquad\qquad W_- f = \lim_{t \to -\infty} U(-t) U_0(t) f.$$

At first glance these definitions do not make sense since the range of $U_0(t)$ is not in general included in the domain of $U(t)$, which is in H. But if the data f have compact support, say f is zero for $|x| > R$, then according to *Huygens' principle* $U_0(t) f$ will be zero in the cone $|x| < |t| - R$ and hence if the obstacle is contained in the ball $\{|x| < \rho\}$, then for such f and for $|t| > R + \rho$, $U_0(t) f$ will belong to H; moreover, for $t > R + \rho$, $U_0(t) f$ satisfies the boundary condition and hence for such t and $s \geq 0$ it follows from the aforementioned uniqueness theorem that $U(s) U_0(t) f$ equals $U_0(s + t) f$, and similarly for $t < -R - \rho$ and $s \leq 0$. This shows that for such an f the right side of (1.2) and (1.3) is independent of t for $t > R + \rho$ and $t < -(R + \rho)$, respectively; therefore, for such f the limits (1.2) and (1.3) trivially exist and $|W_\pm f|_E = |f|_E$. Since data with compact support are dense in H_0, it follows that W_+ and W_- can be extended by continuity to be isometries on all of H_0.

Multiplying (1.2) and (1.3) by $U(s)$ we get

$$(1.4) \quad U(s) W_\pm f = \lim_{t \to \pm\infty} U(s - t) U_0(t - s) U_0(s) f = W_\pm U_0(s) f,$$

so that W_+ and W_- are intertwining transformations for the groups $\{U_0(t)\}$ and $\{U(t)\}$. As a consequence we see that $U(t)$ restricted to the range of W_+ or W_- is unitarily equivalent with $U_0(t)$.

The *scattering operator* S is defined as

$$(1.5) \qquad\qquad S = W_+^{-1} W_-.$$

The operator S is defined on all of H_0 only if the range of W_+ contains the range of W_-, and it defines a unitary operator on H_0 only if the two ranges coincide. It will be shown in Chapter V that for the above wave equation problem the ranges of W_+ and W_- do coincide and are in fact all of H. Assuming this we see by (1.4) that

$$(1.6) \qquad SU_0(t) = W_+^{-1}W_-U_0(t) = W_+^{-1}U(t)W_-$$
$$= U_0(t)W_+^{-1}W_- = U_0(t)S;$$

thus S commutes with $U_0(t)$.

By way of connecting the definition (1.5) of S with the one given in the introduction, we rewrite (1.2) as

$$(1.7) \qquad \lim_{t \to \infty} | U(t)W_+g - U_0(t)g |_E = 0$$

and (1.3) as

$$(1.8) \qquad \lim_{t \to -\infty} | U(t)W_-f - U_0(t)f |_E = 0.$$

Replacing g by Sf in the relation (1.7) and using (1.5) and (1.6) gives

$$(1.9) \qquad \lim_{t \to \infty} | U(t)W_-f - SU_0(t)f |_E = 0.$$

Hence, if we denote $U(t)W_-f$ by u, $U_0(t)f$ by u_- and $SU_0(t)f$ by u_+, then (1.8) asserts that $u - u_-$ tends to zero as t tends to $-\infty$ and (1.9) asserts that $u - u_+$ tends to zero as t tends to $+\infty$.

2. Scattering Theory Formulated in Terms of Representation Theory

The starting point for our approach to scattering theory is a pair of distinguished subspaces D_- and D_+ associated with the group of operators $\{U(t)\}$ relating initial data to data at time t. These are called the *incoming* and *outgoing* subspaces; $D_-[D_+]$ consists of the initial data of those solutions of the wave equation which are zero in a spherical neighborhood $\{|x| < \rho\}$ of the obstacle for all negative [positive] time. We denote the image of $D_\pm$ under $U(t)$ by $D_\pm(t)$.

The following is an immediate consequence of the definitions:

 i. $D_+(t)$ *is a decreasing and* $D_-(t)$ *an increasing family of subspaces; that is, for* $t > s$, $D_+(t)$ *is contained in* $D_+(s)$ *and* $D_-(t)$ *contains* $D_-(s)$.

According to the Holmgren uniqueness theorem (see Theorem 1.5 in Chapter IV), if a solution of the wave equation vanishes in the half cylinder

$\{|x| < \rho, t > 0$ [or $t < 0]\}$, then it vanishes in the entire forward [backward] cone $\{|x| < t + \rho\}$ [or $\{|x| < -t + \rho\}$]. Therefore, every f in $D_+(t)$ [$D_-(t)$] is zero for $|x| < t + \rho$ [$|x| < -t + \rho$]; and this implies:

ii.   *As t tends to $+\infty$ $[-\infty]$, the subspaces $D_+(t)$ $[D_-(t)]$ shrink to the null vector; that is, their intersection is the null vector.*

The next property, proved in Chapter V, lies considerably deeper; in fact it is equivalent to the unitary character of the wave operators introduced in the previous section.

iii.   *As t tends to $-\infty$ $[+\infty]$, the subspaces $D_+(t)$ $[D_-(t)]$ fill out H; that is, their union is dense in H.*

Finally it can be shown (Corollary 2.2, Chapter IV) that:

iv.   *D_- and D_+ are orthogonal to each other.*

Now for any pair of subspaces satisfying the conditions i.–iv. with respect to a group of unitary operators, it is possible to develop an abstract scattering theory. This is accomplished in Chapter II in roughly the following way:

According to a theorem of Ja. G. Sinaĭ (a new proof of which is included in Chapter II), if D_+ is an outgoing subspace for the group $\{U(t)\}$ in the sense of satisfying the conditions i., ii., and iii. above, then H can be represented isometrically as $L_2(-\infty, \infty; N)$, where N is some auxiliary Hilbert space, so that $U(t)$ acts as translation to the right by t units and so that D_+ is represented by $L_2(0, \infty; N)$.† From such a translation representation we can obtain, by Fourier transformation, a spectral representation; according to the Paley–Wiener theorem D_+ is represented here by the space $A_+(N)$ of boundary values of functions analytic in the upper half-plane for which the integrals of their squares are uniformly bounded along horizontal lines.

An analogous representation, *mutatis mutandis*, holds with respect to D_-. We call the representations described above for the group $\{U(t)\}$ and the pair of subspaces D_- and D_+, the *incoming* and *outgoing translation representations;* and we call the representations obtained from these by Fourier transformation the *incoming* and *outgoing spectral representations.*

---

† As we indicate in Section 7, this theorem is equivalent with von Neumann's theorem for operators satisfying the Heisenberg commutation relations.

The auxiliary subspaces N_- and N_+ entering in the incoming and outgoing representations are unitarily equivalent (Theorem 1.2, Chapter II) and therefore can be regarded as identical.

Let $\mathcal{R}_+$ and $\mathcal{R}_-$ denote the operators which assign to a given f in H its outgoing, respectively incoming, translation representer. We call the operator

$$(2.1) \qquad\qquad\qquad S = \mathcal{R}_+\mathcal{R}_-^{-1}$$

the *abstract scattering operator* associated with the group $\{U(t)\}$ and the pair of subspaces D_+ and D_-.

Next we show that the operator S defined as in (2.1) has the following properties:

(a) S *is unitary.*
(b) S *commutes with translation.*
(c) S *maps* $L_2(-\infty, 0; N)$ *into itself.*

Property (a) results from the fact that $\mathcal{R}_+$ and $\mathcal{R}_-$ are both unitary; (b) follows because $\mathcal{R}_+$ and $\mathcal{R}_-$ both define translation representations of the same group of operators. To deduce property (c) we note that every function in $L_2(-\infty, 0; N)$ of the incoming translation representation corresponds to an element f in D_-; since D_- is orthogonal to D_+ we see that the representer of such an f in the outgoing translation representation is orthogonal to $L_2(0, \infty; N)$ and hence belongs to $L_2(-\infty, 0; N)$, as asserted in (c).

Next we go over to the spectral representation by applying the Fourier transformation F. The scattering operator goes into $\mathcal{S} = FSF^{-1}$ and properties (a)–(c) of S imply the following for $\mathcal{S}$:

(a') $\mathcal{S}$ *is unitary.*
(b') $\mathcal{S}$ *commutes with multiplication by scalar functions.*
(c') $\mathcal{S}$ *maps* $A_-(N)$ *into itself.*

According to a special case of a theorem of Fourès and Segal [1], an operator with the properties (a')–(c') can be represented as a multiplicative operator-valued function $\mathcal{S}(\sigma)$, mapping N into N and satisfying the following conditions:

(a'') $\mathcal{S}(\sigma)$ *is the boundary value of an operator-valued function* $\mathcal{S}(z)$ *analytic for* $\operatorname{Im} z < 0$.

(b″) $|S(z)| \leq 1$ *for all z with* $\operatorname{Im} z < 0$.
(c″) $S(\sigma)$ *is unitary for almost all real σ*.

$S(z)$ is the celebrated *Heisenberg scattering matrix*.

In order to see the connection between this abstract approach to scattering and the time-dependent theory sketched in Section 1, we note the following additional properties satisfied by D_- and D_+ :

(α) $U(t)$ *and* $U_0(t)$ *act in the same way on* D_- *for negative t and on* D_+ *for positive t*.

(β) D_- *and* D_+ *are also incoming and outgoing subspaces for* $\{U_0(t)\}$.

(γ)   *The incoming and outgoing translation representations for* $\{U_0(t)\}$ *with respect to* D_- *and* D_+ *coincide, if we map* D_- *onto* $L_2(-\infty, -\rho; N)$ *and* D_+ *onto* $L_2(\rho, \infty; N)$.

Properties (α) and (β) are obvious and (γ) is proved in Section 2, Chapter IV.

First of all these properties enable us to construct the incoming and outgoing translation representations for $\{U(t)\}$ from the corresponding representation for $\{U_0(t)\}$ (which can be described quite explicitly) without appealing to Sinaĭ's theorem. In fact, for f in D_+, simply define k_+ to be equal to the free space representer k, that is the representer of f in the $\{U_0(t)\}$ translation representation; then D_+ maps onto $L_2(\rho, \infty; N)$. Similarly for f in D_- we set k_- equal to k. For any t and f in D_+, we define the representer of $U(t)f$ to be the t-translate of k_+; since by property iii. these elements are dense in H we can, by continuity, define k_+ and k_- for all f in H. The resulting representations are equal to the outgoing and incoming translation representations except for a trivial shift of ρ and $-\rho$, respectively.

Next let $\mathfrak{R}_0$ denote the operator which assigns to each f in H_0 its free space translation representer k; as before, $\mathfrak{R}_+$ and $\mathfrak{R}_-$ will denote the operators assigning to a given f in H its outgoing, respectively incoming, translation representers k_+ and k_- (we ignore the $\pm\rho$ shifts). It follows from the above description of the relation of $k_\pm$ to k that for f in D_+ [or D_-] we have

(2.2) $\mathfrak{R}_+ f = \mathfrak{R}_0 f$ $[\mathfrak{R}_- f = \mathfrak{R}_0 f]$.

Moreover, we see from the definitions (1.2) and (1.3) that for such f

(2.3) $W_+ f = f$ $[W_- f = f]$.

Now, if f has its support in the ball $\{|x| < R\}$, then by Huygens' principle $U_0(R + \rho)f$ belongs to D_+ . Hence, taking (2.3) into account and applying (1.4) we obtain

$$U(R + \rho)W_+ f = W_+ U_0(R + \rho)f = U_0(R + \rho)f .$$

The relation (2.2) now gives

(2.4) $\Re_+ U(R + \rho)W_+ f = \Re_0 U_0(R + \rho)f .$

Since $\Re_+$ and $\Re_0$ are translation representation operators we have

$$\Re_+ U(t) = T(t)\Re_+ \qquad \text{and} \qquad \Re_0 U_0(t) = T(t)\Re_0 ,$$

where $T(t)$ denotes translation to the right by t units. Consequently, (2.4) can be rewritten as

$$T(R + \rho)\Re_+ W_+ f = T(R + \rho)\Re_0 f ;$$

and making use of the fact that $T(t)$ is an isometry we obtain

$$\Re_+ W_+ f = \Re_0 f .$$

This relation extends by continuity to all of H_0 and we conclude that

(2.5) $\Re_+ W_+ = \Re_0 ;$

an analogous argument gives

(2.6) $\Re_- W_- = \Re_0 .$

Referring to definition (1.5) we see that the scattering operator is

$$S = W_+^{-1}W_- = \Re_0^{-1}\Re_+\Re_-^{-1}\Re_0 .$$

Thus the operator $\Re_+\Re_-^{-1}$ defined by (2.1) of this section as the abstract Scattering operator is indeed the scattering operator defined in Section 1, transplanted from H_0 to the real line via the free space translation representation.

In Section 4 we will make use of the following observation: We have seen in (2.2) that $\Re_+ f$ is equal to $\Re_0 f$ for all f in D_+ ; we claim that this implies that $k_+(s) = \Re_+ f$ is equal to $k(s) = \Re_0 f$ for all $s > \rho$ and all f in H.† To prove this we make use of the isometric character of the repre-

† Since f need not lie in H_0 it would be more precise to say that for $s > \rho$ $k_+(s) = \Re_+ f$ is equal to $k(s) = \Re_0 f_0$ for any f_0 in H_0 which coincides with f outside of the ball $\{ |x| > \rho \}$ The rest of the argument is unaffected by this change.

sentations and write

$$(2.7) \qquad\qquad [k, h] = (f, g)_E$$

$$(2.8) \qquad\qquad [k_+, h_+] = (f, g)_E ,$$

where $[\ , \ ]$ denotes the $L_2(-\infty, \infty; N)$ inner product, $(\ , \ )_E$ the inner product associated with the energy norm, g any element of D_+ and $h = \mathfrak{R}_0 g$, $h_+ = \mathfrak{R}_+ g$. Choosing g in D_+ makes $h = h_+$; conversely, by definition of the outgoing representation, every h with support in $s > \rho$ represents some element g of D_+. The equations (2.7) and (2.8) show that $[k - k_+, h] = 0$ for all such h and this implies that $k(s) - k_+(s) = 0$ for $s > \rho$. A similar argument shows that $k_-(s) = \mathfrak{R}_- f$ equals $k(s) = \mathfrak{R}_0 f$ for $s < -\rho$ and all f in H.

We conclude this section by pointing out that the free space translation representation developed in Chapter IV is given by the Radon transform, suitably modified. This leads to an interesting intrinsic characterization of the Radon transform.

3. A Semigroup of Operators Related to the Scattering Matrix

In the dynamic theory of scattering the scattering operator is determined by the behavior of the solution of the perturbed problem at *large distances* for *large negative and positive times*. It turns out, rather surprisingly, that there is also a relation between the behavior of the solution at *small distances* for *large positive times* and the *analytic continuation* of the scattering matrix. This relation can be studied by means of the following set of operators which emphasize the local behavior of the solution:

$$(3.1) \qquad\qquad Z(t) = P_+ U(t) P_- \qquad (t \geq 0) ,$$

where P_+ and P_- are the orthogonal projections onto the orthogonal complements of D_+ and D_-, respectively. Speaking somewhat loosely, the projection operator P_- eliminates the distant past and P_+ eliminates the distant future; $Z(t)$ therefore concentrates on the immediate effect of the scatterer.

It follows from properties i., ii., and iv. of D_- and D_+ that $Z(t)$ (for $t \geq 0$) annihilates D_- and D_+. Let K denote the orthogonal complement of D_- and D_+ :

$$K = H \ominus (D_- \oplus D_-) .$$

Then it can be shown (again for $t \geq 0$) that

(i) $\{Z(t)\}$ *forms a one-parameter semigroup of operators mapping K into K;*

(ii) $Z(t)$ *is a contraction operator, that is,* $|Z(t)|_E \leq 1$;

(iii) $Z(t)$ *tends strongly to zero as t tends to* ∞.

We note that property (iii) above follows from property iii. of $D_\pm$.

Every semigroup of operators satisfying the conditions (i)–(iii) has a translation representation as functions defined on $(-\infty, 0]$, obtained by assigning to each f in K the function

$$h(s) = Z(-s)f \qquad (-\infty < s \leq 0).$$

The action of $Z(t)$ in this representation is clearly a shift to the right by t units followed by restriction to the negative axis. Next let B denote the infinitesimal generator of the semigroup $\{Z(t)\}$ and define the N-norm for the vectors in K as

$$|f|_N^2 = -2\,\mathrm{Re}\,(f, Bf)_K.$$

Then, as shown in Section 2, Chapter III, the K-norm, of f and the $L_2(-\infty, 0; N)$ norm of h are equal.

It is clear that the subspace of functions in $L_2(-\infty, 0; N)$ representing K is invariant under right translation followed by restriction to $(-\infty, 0]$; consequently its orthogonal complement with respect to $L_2(-\infty, 0; N)$ will be invariant under translation to the left. According to a generalization of a theorem of Beurling (see Lax [2, 3], Halmos [1]), the Fourier transform of such a left invariant subspace can be represented, essentially uniquely, as

$$\mathcal{S}'(z)A_-(N).$$

As before $A_-(N)$, the Fourier transform of $L_2(-\infty, 0; N)$, is the space of boundary values of functions analytic in the lower half-plane and square integrable along the lines $\mathrm{Im}\, z = c < 0$. $\mathcal{S}'(z)$ is an *operator-valued inner factor*; that is, $\mathcal{S}'(z)$ is analytic of norm ≤ 1 in the lower half-plane, and its values on the real axis are isometries. We call this inner factor $\mathcal{S}'$ *the scattering matrix associated with the semigroup* $\{Z(t)\}$.

In case $\{Z(t)\}$ is the semigroup defined by (3.1) in terms of a unitary group $\{U(t)\}$ and an orthogonal pair of incoming and outgoing subspaces, the translation representation described above for $\{Z(t)\}$ is the same as the outgoing translation representation for $\{U(t)\}$ constructed in Section 2. In the latter representation the subspace of functions representing K is $L_2(-\infty, 0; N) \ominus SL_2(-\infty, 0; N)$, that is the orthogonal complement with respect to $L_2(-\infty, 0; N)$ of $SL_2(-\infty, 0; N)$. According to the definition given in Section 2 for the scattering matrix, the Fourier transform of this subspace is $S(z)A_-(N)$. Since by properties (a'), (b'), (c') derived in Section 2, the scattering matrix is also an inner factor, it follows by uniqueness that the scattering matrix associated with the group $\{U(t)\}$ and $D_\pm$ is the same as the scattering matrix associated with the semigroup (3.1).

Next we establish an important connection between the spectrum of $\{Z(t)\}$ and the points z in the lower half-plane at which the operator $S(z)$ is singular. Suppose that μ is an eigenvalue of the infinitesimal generator B of $\{Z(t)\}$; then the corresponding eigenfunction can be represented in $L_2(-\infty, 0; N)$ by an exponential function:

$$h(s) = e^{-\mu s}n \qquad (s \le 0) ,$$

where n belongs to N. As noted above, h represents an element of K if and only if it is orthogonal to $SL_2(-\infty, 0; N)$; that is

$$[Sk, h] = 0$$

for all k in $L_2(-\infty, 0; N)$. By Parseval's formula this can be written in terms of the Fourier transforms as

$$[\hat{Sk}, \hat{h}] = 0 ;$$

and since

$$\hat{h}(\sigma) = \frac{n}{i\sigma - \mu}$$

the above condition amounts to

$$\int_{-\infty}^{\infty} \frac{(S(\sigma)\hat{k}(\sigma), n)_N}{i\sigma + \bar{\mu}} \, d\sigma = 0$$

for all $\hat{k}$ in $A_-(N)$. Now both S and $\hat{k}$ are regular in the lower half-plane, $S(z)$ being bounded and $\hat{k}(z)$ being uniformly square integrable along all lines $\mathrm{Im}\, z = c < 0$; hence this integral equals the residue at $\sigma = i\bar{\mu}$, and we obtain

$$(S(i\bar{\mu})\hat{k}(i\bar{\mu}), n)_N = 0 .$$

Since the value of $\hat{k}$ at $i\bar{\mu}$ can be any vector in N, it follows that μ *belongs to the point spectrum of B if and only if the range of S at $z = i\bar{\mu}$ has positive codimension.* More generally we show in Chapter III that if μ belongs to the spectrum of B then $S(i\bar{\mu})$ is a singular operator, and conversely.

From this close connection between the spectrum of the semigroup and the points in the lower half-plane at which the scattering matrix $S(z)$ is singular we can by studying either one, get information about the other. In the case of the wave equation in an exterior domain the semigroup can be examined directly with methods from the theory of hyperbolic partial differential equations. For instance, it is shown in Chapter V that there exist positive T and κ for which the operator $Z(T)(\kappa I - B)^{-1}$ is compact; from this it follows, as shown in Chapter III, that B has a pure point spectrum contained in the half-plane $\operatorname{Re}\mu < 0$; and this in turn implies that $S(z)$ is invertible at all but a discrete set of points in the lower half-plane. Since $S(z)$ is unitary on the real axis, it can then be continued analytically to the upper half-plane by the Schwarz reflection principle with

$$S(z) = [S^*(\bar{z})]^{-1}$$

for $\operatorname{Im} z > 0$. The scattering operator is then meromorphic in the whole plane having as its poles the points z for which iz lies in the spectrum of B.

We also show in Chapter V that if the scattering object is star-shaped then for some positive T, $|Z(T)| < \alpha < 1$. This implies that the spectrum of B lies to the left of the line $\operatorname{Re} z = (\log \alpha)/T$ and that the scattering matrix has a bounded holomorphic extension into the strip $\operatorname{Im} z < -(\log \alpha)/T$.

In case $Z(T)$ is compact for some T it can be shown that the eigenfunction expansion of $\{Z(t)\}$ is asymptotically valid for large t. This furnishes us with a precise relation between the poles of the scattering matrix and the local behavior of the solution for large positive times. As a corollary we also obtain the fact that when $Z(T)$ is compact for some T, the local energy decays exponentially.

We conclude this section by establishing a connection between the eigenfunctions of $\{Z(t)\}$ and certain highly improper eigenfunctions of $\{U(t)\}$: For any positive a we define the subspaces $D_{\pm}{}^a$ by

$$D_-{}^a = U(-a)D_- \qquad \text{and} \qquad D_+{}^a = U(a)D_+ .$$

Since D_- and D_+ are an orthogonal pair of incoming and outgoing subspaces, so are $D_-{}^a$ and $D_+{}^a$ for positive a. Therefore we can associate a scattering matrix $\mathcal{S}^a(z)$ with this new pair of subspaces; it is easy to prove that $\mathcal{S}^a$ depends only in an inessential fashion on a:

$$\mathcal{S}^a(z) = \exp(-i2az)\mathcal{S}(z) .$$

Let $\{Z^a(t)\}$ denote the semigroup of operators corresponding to the pair $D_\pm{}^a$ and let B^a be the infinitesimal generator of $\{Z^a(t)\}$. Since the location of the singularities of $\mathcal{S}^a$ clearly does not depend on a it follows from the foregoing that neither does the location of the spectrum of B^a. In fact it is not difficult to show for $a < b$ that $P_+{}^a$, the projection onto the orthogonal complement of $D_+{}^a$, maps the null space of $\mu I - B^b$ onto that of $\mu I - B^a$ in a one-to-one fashion:

$$f^a = P_+{}^a f^b .$$

We show in Chapter V that as b tends to infinity the inductive limit of f^b exists as an improper eigenfunction of $\{U(t)\}$. This gives still another, and very concrete, way of characterizing the poles of the scattering matrix.

4. The Form of the Scattering Matrix

The theories outlined in the foregoing sections give a clearcut definition of the scattering operator and yield a lot of useful information about the analytic properties of the scattering matrix; what is lacking however is a method for actually calculating it. In this section we derive, somewhat heuristically, an expression for the scattering matrix in the form of the identity plus an integral operator whose kernel is the so-called *transmission coefficient*. The transmission coefficient is defined in terms of the asymptotic value near infinity of the *scattered wave*, which itself is defined as the solution of the *reduced wave equation* in the exterior of the obstacle whose values on the obstacle are prescribed to be the same as those of a plane wave and which satisfies the *outgoing Sommerfeld condition*.

All of these assertions will be proved rigorously in Chapter V; the key theorem on the existence of the scattered wave solutions being deduced from the fact that the spectrum of the infinitesimal generator of the semigroup defined in Section 3 contains no point on the imaginary axis.

As defined in Section 2 the scattering operator S relates the incoming and outgoing translation representations. S commutes with translation and hence is a convolution operator:

$$(4.1) \qquad k_+(s) = [Sk_-](s) = \int_{-\infty}^{\infty} S(r)k_-(s-r)\, dr.$$

The Fourier transform of $S(r)$ has been defined as the scattering matrix $\mathbb{S}(\sigma)$. Therefore, if we choose $k_-(s)$ to be $e^{-i\sigma s}n$ with n in N (actually this is not allowable since the scattering operator acts only on square integrable functions), then we get

$$(4.2) \qquad S(e^{-i\sigma s}n) = e^{-i\sigma s}\mathbb{S}(\sigma)n.$$

Now, what f is represented by $k_- = e^{-i\sigma s}n$? Translation to the right by t units merely multiplies k_- by $e^{i\sigma t}$; hence the action of $U(t)$ on f also amounts to multiplication by $e^{i\sigma t}$; in other words the solution u with initial data f depends exponentially on t:

$$u(x, t) = e^{i\sigma t}v(x).$$

Thus the initial data f for u are

$$(4.3) \qquad f = \{v, i\sigma v\}.$$

Since u satisfies the wave equation, v satisfies the reduced wave equation

$$(4.4) \qquad \Delta v + \sigma^2 v = 0,$$

and since u vanishes on the boundary of the obstacle, so does v:

$$(4.5) \qquad v = 0 \qquad \text{on the boundary}.$$

Conversely, if f is of the form (4.3) and v satisfies (4.4) and (4.5), then $v(t)f$ depends exponentially on t and therefore both $\mathcal{R}_- f$ and $\mathcal{R}_+ f$ are exponential functions of s.

As remarked at the end of Section 2, the incoming translation representation $\mathcal{R}_- f$ of f equals its free space translation representation $\mathcal{R}_0 f$ for $s < -\rho$. Since $\mathcal{R}_- f = e^{-i\sigma s}n$ we see that

$$\mathcal{R}_0 f = e^{-i\sigma s}n \qquad \text{for} \quad s < -\rho.$$

Next define f_0 as that element of H_0 for which

$$\mathcal{R}_0 f_0 = e^{-i\sigma s}n \qquad \text{for all} \quad s.$$

By the same analysis as before we see that f_0 is of the form $\{v_0, i\sigma v_0\}$ where v_0 satisfies the reduced wave equation for all x but of course does not vanish on the boundary of the obstacle. Thus f_0 differs from the f we are looking for by a correction term p with the following properties:

(i) p is of the form $\{w, i\sigma w\}$;

(ii) w satisfies the reduced wave equation in the exterior domain and equals v_0 on the boundary;

(iii) $\mathcal{R}_0 p = 0$ for $s < -\rho$.

We will establish the existence of such a p in Chapter V; this gives us the desired function

$$(4.6) \qquad\qquad f = f_0 - p .$$

Our next task is to determine the outgoing representation of f. According to the observation made at the end of Section 2, $\mathcal{R}_+ f = R_0 f$ for $s > \rho$; that is

$$(4.7) \quad \mathcal{R}_+ f = \mathcal{R}_0 f = \mathcal{R}_0 f_0 - \mathcal{R}_0 p = e^{-i\sigma s} n - \mathcal{R}_0 p \qquad \text{for} \quad s > \rho .$$

Since $\mathcal{R}_+ f$ is an exponential function of s it follows from (4.7) that $R_0 p$ is an exponential function of s for $s > \rho$:

$$(4.8) \qquad\qquad \mathcal{R}_0 p = e^{-i\sigma s} m \qquad \text{for} \quad s > \rho .$$

Thus $R_0 f = e^{-i\sigma s}(n - m)$ for $s > \rho$ and this together with the above noted uniqueness allows us to identify f with the function whose outgoing translation representation is $e^{-i\sigma s}(n - m)$ for all s:

$$(4.9) \qquad\qquad \mathcal{R}_+ f = e^{-i\sigma s}(n - m) \qquad \text{for all} \quad s .$$

Finally combining (4.1), (4.2), and (4.9) we obtain the relation,

$$S(\sigma)n = n - m .$$

If we now denote by $\mathfrak{I}(\sigma)$ the operator relating n to m, we can then write this as

$$S(\sigma) = I - \mathfrak{I}(\sigma) .$$

In order to get a more specific description of the operator $\mathfrak{I}$ we have to use more specific facts about the free space translation representation. In Chapter IV we show that:

(a)   The elements of the auxiliary space N are square integrable functions on the two-dimensional unit sphere, S_2 .

(b)   If p in H_0 is of the form $\{w, i\sigma w\}$ where w is a solution of the reduced wave equation (4.4), and if the free space translation representation of p vanishes for $s < -\rho$, then w satisfies the outgoing Sommerfeld radiation condition,† that is, for r large

$$(4.10) \qquad\qquad w(x) \sim \frac{e^{-i\sigma r}}{r}\, m(\omega), \qquad x = r\omega,$$

where m is the same function of ω that enters the translation representation (4.8).

We show in Chapter V that the operator $\mathfrak{I}(\sigma)$ is an integral operator with a smooth, in fact analytic, kernel. Taking this for granted, the kernel $t(\omega, \theta; \sigma)$ of $\mathfrak{I}(\sigma)$ equals $[\mathfrak{I}\, \delta_\theta](\omega)$, δ_θ being the delta-function whose support consists of the point θ. An easy explicit calculation shows that the f_0 in H_0 represented by $e^{-i\sigma s}\, \delta_\theta$ is the initial data of the *plane wave solution* $u(x, t) = \exp[i\sigma(t - x \cdot \theta)]$; the corresponding correction term p is the scattered wave. This verifies our previous description of the scattering matrix in terms of the asymptotic values near infinity of solutions of the reduced wave equation with given boundary values on the obstacle and satisfying the radiation condition.

5.   A Simple Example

We shall illustrate the general theory by a simple one-dimensional example. The unperturbed system satisfies the wave equation

$$(5.1) \qquad\qquad u_{tt} - u_{xx} = 0$$

on the positive half-line $x > 0$, with the boundary condition

$$(5.2) \qquad\qquad u(0, t) = 0$$

at $x = 0$, and initial data

$$(5.3) \qquad\qquad u(x, 0) = f_1(x), \qquad u_t(x, 0) = f_2(x)$$

prescribed at $t = 0$. Multiplying (5.1) by u_t and integrating by parts, we

---

† In fact we prove that the following three conditions are equivalent: (a) The free space translation representation of p vanishes for $s < -\rho$; (b) $U_0(t)p$ vanishes for $|x| < t - \rho$; and (c) w behaves asymptotically like (4.10) for large $|x|$.

see that the total energy E, defined as

(5.4)
$$E = \frac{1}{2} \int_0^\infty [u_x^2 + u_t^2]\, dx,$$

is conserved; that is, E is independent of time. Since E is a positive definite functional for functions satisfying the boundary condition (5.2), it follows that the solutions of the unperturbed problem are uniquely determined by the initial data. We denote the set of all initial data of finite energy by H_0.

This problem can be solved explicitly for arbitrary f_1 and f_2 by writing u in the form of the general solution of the wave equation:

(5.5)
$$u(x, t) = a(t + x) + b(t - x)$$

and choosing a and b to fit the initial and boundary conditions. The initial data require that

(5.6)
$$a(x) + b(-x) = f_1(x) \qquad (x > 0),$$
$$a'(x) + b'(-x) = f_2(x) \qquad (x > 0),$$

and this determines the function a for positive arguments and b for negative arguments, at least up to a constant which we are free to adjust so that $a(0) = 0 = b(0)$. This is consistent with the boundary condition requirement:

(5.7)
$$a(t) + b(t) = 0,$$

which then determines the function a for negative arguments and b for positive arguments.

We denote by $U_0(t)$ the operator relating the initial data $\{f_1, f_2\}$ to the solution data at time t.

Assertion *The mapping*

(5.8)
$$\{f_1, f_2\} \to h(s) = a'(-s) \qquad in \quad L_2(-\infty, \infty)$$

defines a translation representation for the group $\{U_0(t)\}$ which is both incoming and outgoing, and unitary.

Proof. Let u_T denote the shifted function

$$u_T(x, t) = u(x, t + T).$$

According to (5.5)

$$u_T(x, t) = a_T(t + x) + b_T(t - x)$$

where $a_T(s) = a(s + T) - a(T)$ and $b_T(s) = b(s + T) - b(T)$. This proves that (5.8) is a translation representation.

To prove that it is outgoing, suppose that $h(s)$ is zero for all $s < 0$; then by (5.8) and the condition $a(0) = 0$, $a(s)$ is zero for $s > 0$, and by (5.7) so is $b(s)$. Formula (5.5) then shows that u vanishes for $0 < x < t$; that is, u is outgoing. Conversely, it follows easily from (5.5) that if u is outgoing then a and b are zero for positive arguments. Thus the translation representation (5.8) is indeed outgoing; and it can be shown in a similar fashion that it is also incoming.

Using the definition (5.4) of the energy and the relation (5.6) we obtain

$$(5.9) \quad E = \frac{1}{2} \int_0^\infty \left[(f_1')^2 + (f_2)^2 \right] dx$$

$$= \frac{1}{2} \int_0^\infty \left[(a'(x) - b'(-x))^2 + (a'(x) + b'(-x))^2 \right] dx$$

$$= \int_0^\infty \left[(a'(x))^2 + (b'(-x))^2 \right] dx.$$

By (5.7) $b'(-x) = -a'(-x)$ so that (5.9) can be rewritten as

$$E = \int_0^\infty \left[(a'(x))^2 + (a'(-x))^2 \right] dx = \int_{-\infty}^\infty (a'(x))^2 \, dx;$$

this shows that the representation (5.8) is isometric. Conversely, given any function $a(s)$ with square integrable first derivative and vanishing at $s = 0$, we define b by (5.7) and f_1 and f_2 by (5.6); then $f = \{ f_1, f_2 \}$ is represented by $h(s) = a'(-s)$. This proves that the representation is unitary.

We now perturb the problem just considered by replacing the boundary condition (5.2) with the following:

$$(5.2)' \qquad\qquad u_x(0, t) = u(0, t).$$

In this case the integration by parts technique leads to the following expression for the total energy:

$$(5.4)' \qquad E = \frac{1}{2} \int_0^\infty [u_x{}^2 + u_t{}^2]\, dx + \tfrac{1}{2}(u(0))^2.$$

Again, E is conserved in time and since it is a positive definite functional on the solutions (5.1) with (5.2)', the solutions are uniquely determined by their initial data. We denote by $U(t)$ the operator relating initial data to the data at time t.

The problem (5.1), (5.2)', (5.3) can also be solved explicitly in the same way as before: Write u in the form (5.5); the initial data again determine, by means of (5.6), the functions a and b for positive, respectively negative, arguments, up to one constant. The boundary condition

$$(5.7)' \qquad a'(t) - b'(t) = a(t) + b(t)$$

is now in the form of an ordinary differential equation which can be used to determine a for negative arguments and b for positive arguments. Conversely, given any function $a(t)$ defined for all real t and whose first derivative is square integrable, then as is easily proved, there exists *exactly one* function $b(t)$ defined on the whole real axis with square integrable first derivative, which satisfies (5.7)', namely

$$b(t) = -a(t) + 2e^{-t} \int_{-\infty}^t e^s a'(s)\, ds.$$

If we define $u(x, t)$ by (5.5), then u will be a weak solution of the wave equation, with finite energy, satisfying the boundary condition (5.2)' in a weak sense.

Assertion   *The mappings*

$$h_+(s) = b'(-s)$$

$$(5.8)' \qquad \{f_1, f_2\} \begin{matrix} \nearrow \\ \\ \searrow \end{matrix}$$

$$h_-(s) = a'(-s)$$

define translation representations of $\{U(t)\}$*, outgoing and incoming respectively, both of which are unitary.*

Proof. It follows as in the unperturbed case that (5.8)' determines translation representations. To show that the representation given by

$b'(-s)$ is outgoing, consider an outgoing solution $u(x, t)$; that is, a solution for which $u(x, t)$ vanishes for $0 < x < t$. As before both functions a and b entering the representation (5.5) of u are easily proved from this representation to be constant for positive arguments. Conversely, suppose that $b(s)$ is constant for positive arguments. Then it follows from (5.7)′ that for positive s the function a satisfies the equation

$$a' - a = b(0) .$$

Since each solution to this equation is equal to $-b(0) + ce^s$ and since a' is required to be square integrable, we see that $a(s) = -b(0)$ for $s > 0$. But then $u(x, t)$ given by (5.5) vanishes for $0 < x < t$; this shows that outgoing solutions correspond to those b' which are zero for $s > 0$; that is, to those $h_+(s)$ which are zero for $s < 0$. The incoming character of the representation $f \rightarrow h_-$ can be established similarly.

In order to prove that these representations are unitary we use the definition (5.4)′ of energy and (5.6):

$$(5.9)' \quad E = \frac{1}{2} \int_0^\infty \left[(f_1')^2 + (f_2)^2 \right] dx + \tfrac{1}{2} (f_1(0))^2$$

$$= \frac{1}{2} \int_0^\infty \left[(a'(x) - b'(-x))^2 + (a'(x) + b'(-x))^2 \right] dx$$

$$+ \tfrac{1}{2} (a(0) + b(0))^2$$

$$= \int_0^\infty (a'(x))^2 \, dx + \int_{-\infty}^0 (b'(x))^2 \, dx + \tfrac{1}{2} (a(0) + b(0))^2.$$

Multiplying the boundary condition (5.7)′ by $a' + b'$, we get

$$(5.10) \qquad\qquad (a')^2 - (b')^2 = (a + b)(a + b)';$$

and integrating from 0 to ∞ gives

$$\int_0^\infty (a'(x))^2 \, dx - \int_0^\infty (b'(x))^2 \, dx + \tfrac{1}{2} (a(0) + b(0))^2 = 0.$$

Subtracting this from the right side of (5.9)′ we finally obtain

$$(5.11)_b \qquad\qquad E = \int_{-\infty}^\infty (b'(x))^2 \, dx.$$

On the other hand, integrating (5.10) from $-\infty$ to 0 gives

$$\int_{-\infty}^{0} (a'(x))^2\, dx - \int_{-\infty}^{0} (b'(x))^2\, dx - \tfrac{1}{2}(a(0) + b(0))^2 = 0;$$

and adding this to (5.9)′ yields

(5.11)ₐ
$$E = \int_{-\infty}^{\infty} (a'(x))^2\, dx.$$

The relations (5.11)ₐ and (5.11)ᵦ establish the isometric character of the mappings (5.8)′. Since b [or a] can be chosen arbitrarily as long as its first derivative is square integrable the h_+ [respectively, h_-] representation is unitary.

As we have indicated in Section 2 the scattering operator maps h_- into h_+. To obtain the corresponding scattering matrix S we first take the Fourier transform of (5.7)′; denoting the dual variable by σ we get

$$-i\sigma(\hat{a} - \hat{b}) = \hat{a} + \hat{b},$$

which gives

(5.12)
$$\hat{b}(\sigma) = \frac{i\sigma + 1}{i\sigma - 1}\,\hat{a}(\sigma).$$

Taking the Fourier transform of (5.8)′ we obtain

$$i\sigma\hat{b}(-\sigma) = \hat{h}_+(\sigma) \qquad \text{and} \qquad i\sigma\hat{a}(-\sigma) = \hat{h}_-(\sigma);$$

and substituting these into (5.12) yields

(5.13)
$$\hat{h}_+(\sigma) = -\,\frac{1 - i\sigma}{1 + i\sigma}\,\hat{h}_-(\sigma).$$

This shows that the scattering matrix† (in this case an one by one matrix) is

(5.14)
$$S(\sigma) = -\,\frac{1 - i\sigma}{1 + i\sigma}$$

† The minus sign in formula (5.14) is of no real significance; it appears because our choice in (5.8)′ of outgoing representers of elements in D_+ is the negative of their unperturbed representers given by (5.8) rather than being equal to their unperturbed representers as in Section 2.

which is indeed of modulus one on the real axis and meromorphic in the whole plane, having a single pole located in the upper half-plane at $\sigma = i$.

We now construct the semigroup $\{Z(t)\}$; for this purpose we first determine the orthogonal complement K of D_- and D_+. Because of the fact that we were able to construct a translation representation in the unperturbed case which was literally both incoming and outgoing, it follows that D_- and D_+ together span the set of all initial data in H_0. The perturbed energy differs from the unperturbed energy by the term $\frac{1}{2}(f_1(0))^2$; since the boundary conditions for the unperturbed problem demand that f_1 vanish at $x = 0$, it follows that H_0 is isometrically imbedded in H, the space of initial data for the perturbed problem. Furthermore, H can be split as the direct sum of H_0 and the one-dimensional space of data spanned by $\{1, 0\}$:

$$H = H_0 \oplus \{c\{1, 0\}\} .$$

Clearly, this is an orthogonal decomposition; and since H_0 is $D_- \oplus D_+$ we see that K, the orthogonal complement of $D_- \oplus D_+$, is the one-dimensional space spanned by $\{1, 0\}$. The projection of any f in H into K is simply

(5.15) $$Pf = \{f_1(0), 0\} .$$

Next we solve the perturbed initial value problem for initial data in K:

$$u(x, 0) = 1 , \qquad u_t(x, 0) = 0 \qquad \text{for} \quad x \geq 0 .$$

In the domain of dependence of the initial interval $(0, \infty)$, the boundary condition does not influence the solution at all and hence u is equal to one there:

$$u(x, t) = 1 \qquad \text{for} \quad x \geq |t| .$$

In order to determine u in the triangular region between the positive t-axis and the line $x = t$, we write $u(x, t)$ in the form $a(t + x) + b(t - x)$; setting $x = t$ and using the fact that $u(x, x) = 1$ for all $x > 0$, we conclude that $a(x)$ is constant for $x > 0$ and thus may be absorbed into $b(x)$, again for $x > 0$. The function b is then determined by the boundary condition (5.2)$'$:

$$-b'(t) = u_x(0, t) = u(0, t) = b(t) \qquad \text{for} \quad t > 0 ,$$

and

$$b(0) = u(0, 0) = 1 .$$

Consequently, $b(t) = e^{-t}$ for $t > 0$ and hence

$$u(x, t) = e^{x-t} \qquad \text{for} \quad 0 < x < t .$$

Similarly,

$$u(x, t) = e^{t-x} \qquad \text{for} \quad 0 < x < -t .$$

Since $Z(t)$ was defined as $PU(t)$ restricted to K, we find from the formula (5.15), describing the action of P, that

$$Z(t)\{1, 0\} = e^{-t}\{1, 0\} \qquad (t \geq 0) .$$

Thus, $f^0 = \{1, 0\}$ is an eigenfunction of $Z(t)$ with eigenvalue e^{-t}. The corresponding eigenvalue of the infinitesimal generator B is -1. Since $\sigma = -i$ is a zero of the scattering matrix $S(\sigma)$ given by (5.14), this bears out the assertion in Section 3 about the relation between the spectrum of B and the zeros of the scattering matrix.

Next we characterize the shifted subspaces $D_+{}^c = U(c)D_+$ and $D_-{}^c = U(-c)D_-$, c any positive number. This is easy to do in terms of the translation representation $(5.8)'$: $D_-{}^c$, $[D_+{}^c]$ consists of data represented by functions a $[b]$ whose derivatives are zero for $s < c$ [respectively $s > -c$]. If a and b correspond to data f orthogonal to both $D_-{}^c$ and $D_+{}^c$, then a' vanishes for $s > c$ and b' vanishes for $s < -c$. In view of the relation (5.6) between a, b, and the data, we see that for such f the first component f_1 is constant and the second component f_2 is zero for $x > c$. This characterizes K^c; the projection P^c of f into K^c is given by

$$P^c f = f(x) \qquad (0 < x < c) ,$$

$$= \{ f_1(c), 0 \} \qquad (c \leq x) .$$

According to the general theory developed in Section 3, the semigroup operator $Z^c(t) = P^c U(t)$, restricted to K^c, has an eigenfunction f^c with eigenvalue e^{-t}. We now determine f^c: Let u^c denote the solution to the perturbed problem with initial data f^c; since the action of P^c leaves the data unaltered for $x < c$ we may write

$$U(t)f^c = Z^c(t)f^c = e^{-t}f^c \qquad \text{for} \quad x < c .$$

This shows that for $x < c$, u^c is of the form

$$u^c(x, t) = e^{-t}v(x) \qquad (x < c) .$$

Since u^c is a solution of the wave equation, v must be of the form $\alpha e^x + \beta e^{-x}$; and since v also satisfies the boundary condition $(5.2)'$: $v_x(0) = v(0)$, β must be equal to zero. Consequently,

$$u^c(x, t) = e^{x-t} \qquad (x < c) .$$

Clearly, u^c is the restriction to $x < c$ of

$$u(x, t) = e^{x-t} ,$$

the projective limit of u^c as c tends to infinity. The initial data of u are $\{e^x, -e^x\}$; this is an eigenvector of $U(t)$ and since it is exponentially increasing it is a highly improper one.

    The form of the scattering matrix will now be rederived by the method described in Section 4 involving steady state solutions of the wave equation:

$$u(x, t) = e^{i\sigma t}v(x) .$$

Thus, v is a solution of the reduced wave equation

$$v_{xx} + \sigma^2 v = 0$$

and is therefore of the form

$$(5.16) \qquad\qquad v(x) = \alpha e^{i\sigma x} + \beta e^{-i\sigma x} .$$

The unperturbed boundary condition (5.2) is satisfied if

$$(5.17) \qquad\qquad \beta = -\alpha ,$$

while the perturbed boundary condition $(5.2)'$ is satisfied if

$$i\sigma(\alpha - \beta) = \alpha + \beta ,$$

that is if

$$(5.17)' \qquad\qquad \beta = \frac{i\sigma - 1}{i\sigma + 1}\, \alpha .$$

By (5.16)

$$u(x, t) = \exp{(i\sigma t)}v = \alpha \exp{\left[(i\sigma(t + x)\right]} + \beta \exp{\left[i\sigma(t - x)\right]} .$$

The first term represents an incoming wave, the second an outgoing wave which can be regarded as the reflection of the incoming wave. Thus the effect of the perturbation of the boundary condition is to change the coefficient of the reflected wave by the factor

$$\frac{1 - i\sigma}{1 + i\sigma} \; ;$$

this is precisely the scattering matrix.

6. Scattering Theory for Transport Phenomena

The streaming of particles reflected by an obstacle in an n-dimensional real Euclidean space furnishes us with a different and in some ways more transparent application of our general theory. We assume that the particles are reflected according to the usual laws of reflection: the incident and reflected rays lie in one plane with the normal to the obstacle and make equal angles with the normal. In this section we show, somewhat sketchily, how this phenomena fits into the framework described in sections two and three; for details see Lax and Phillips [5].

We denote the domain exterior to the obstacle by G, the boundary of the obstacle by ∂G, and assume that ∂G is bounded and twice continuously differentiable. All particles are assumed to travel with unit speed; therefore the appropriate phase space Ω_0 is the Cartesian product

$$\Omega_0 = G \times S_{n-1} \,,$$

where S_{n-1} is the unit sphere in R_n. Finally, we denote the elements of Ω_0 by $\omega = \{x, \theta\}$.

The trajectory $\tau_t(\omega)$ of a particle situated at the point ω is a straight line between reflections:

$$\tau_t(\omega) = \{x + t\theta, \theta\} \; ;$$

here $\omega = \{x, \theta\}$. At a point of reflection a trajectory is discontinuous; and if a trajectory is tangent at some point to the obstacle then past such a point, the first derivative of $\tau_t(\omega)$ can be discontinuous in ω. It is easy to show that for fixed t all the discontinuities of $\tau_t(\omega)$ lie on a piecewise smooth hypersurface in Ω_0 and that away from these discontinuities $\tau_t(\omega)$ is a differentiable function of ω and t.

We impose as measure m on Ω_0 the product of the Lebesgue measures of G and S_{n-1}. The following extension of Liouville's theorem is not hard to prove:

Theorem 6.1.   *The flow τ_t on Ω_0 is measure preserving; that is for every t and every measurable set Σ*

$$m(\Sigma) = m(\tau_t(\Sigma)) .$$

In what follows we shall denote the image of any set Σ under τ_t by $\Sigma(t)$. A point ω is called *free forward* if the trajectory through ω has no points of reflection for t large enough; *free backward* is defined similarly. A point not free forward is called *trapped forward*; and *trapped backward* is defined similarly. If a point is trapped both forward and backward we simply call it *trapped*. Trapped points play no role in scattering; we shall ignore them and deal exclusively with their complement, denoted by Ω.

Theorem 6.2.   *The set of points trapped in one direction but not in the other has measure zero.*

Proof.   Denote by Ω_+ the set of points trapped backward but not forward, by Ω_- those trapped forward but not backward; then $\Omega = \Omega_+ \cup \Omega_-$. By definition, for any ω in Ω_+ there exists a largest time T such that $\tau_T(\omega)$ is a point of reflection; on the other hand there are arbitrarily large negative times for which $\tau_t(\omega)$ is a point of reflection. It follows from this that for $t < T$ the spatial part of every point of the trajectory $\tau_t(\omega)$ lies on a line segment whose end points belong to ∂G; thus all of these spatial parts lie in the convex hull of ∂G. Since ∂G was assumed to be bounded all points $\tau_t(\omega)$ with $t < T$ lie in the bounded set B consisting of the product of the convex hull of ∂G and S_{n-1}.

Next, let j be any integer and denote by Ω_j the set of all those points ω in Ω_+ for which T lies between j and $j + 1$. Clearly

(6.1) $\Omega_+ = \cup \, \Omega_j$;

therefore, if we show that all sets Ω_j have measure zero it will follow by countable additivity that Ω_+ is also of measure zero.

By construction, the set $\Omega_j(t)$ into which Ω_j is carried by the flow after time t belongs to B for $t \le j$. In particular $\Omega_j(j)$ is a subset of B and it follows then that $\Omega_j(j + 1)$ is contained in $B(1)$ so that it is also bounded. Let d denote an upper bound on the diameter of $\Omega_j(j + 1)$; that is the distance between the spatial parts of any two points of $\Omega_j(j + 1)$ is less than d. Now by definition, for $t > j + 1$ each point of $\Omega_j(t)$ is obtained by carrying the corresponding point of $\Omega_j(j + 1)$ along a straight line for a distance of $t - (j + 1)$. Since the distance between the spatial parts of any two points ω and ω' of $\Omega_j(j + 1)$ is less than d, it follows that if t and t' differ by at least d and if both are greater than $j + 1$, then $\tau_t(\omega)$ and $\tau_{t'}(\omega')$ are different; this means that the sets $\Omega_j(t)$ and $\Omega_j(t')$ are disjoint. In particular for any positive integer k the sets

$$(6.2) \quad \Omega_j(j + 1 + d), \quad \Omega_j(j + 1 + 2d), \quad \cdots, \quad \Omega_j(j + 1 + kd)$$

are pairwise disjoint. Since the flow is measure preserving, each of the sets (6.2) has the same measure, say m, as the set Ω_j, and since they are disjoint their union has measure km.

On the other hand since $\Omega_j(t)$ lies in B for $t < j$, it follows that each point of the sets (6.2) is carried into B by the flow after time $-(1 + kd)$. Again, using the measure preserving character of the flow it follows that the measure of B is greater than km. Since B, being bounded, has finite measure and since k was an arbitrary positive integer, this can be so only if the measure m of Ω_j is zero. Combining this with (6.1) we see that $m(\Omega_+) = 0$. That $m(\Omega_-) = 0$ can be proved similarly; this completes the proof of Theorem 6.2.

Let ρ denote the radius of a ball around the origin which contains all points of ∂G. The point sets E_+ and E_- defined as

$$(6.3)_+ \qquad\qquad E_+ = [\{x, \theta\} ; \quad x \cdot \theta \geq \rho],$$

$$(6.3)_- \qquad\qquad E_- = [\{x, \theta\} ; \quad x \cdot \theta \leq -\rho]$$

certainly have the property that E_+ is free forward and E_- is free backward. Further properties are summarized in

Theorem 6.3.

$(i)_+ \quad E_+(t) \subset E_+ \qquad for \quad t \geq 0$

$(i)_- \quad E_-(t) \subset E_- \qquad for \quad t \leq 0$

$(ii)_+ \quad \cap E_+(t) = \phi ,$

$(ii)_- \quad \cap E_-(t) = \phi ,$

$(iii) \quad Both\ sets$

$$\cup E_+(t) \qquad and \qquad \cup E_-(t)$$

are subsets of Ω which differ from Ω by a set of measure zero,

$(iv) \quad E_+ \cap E_- = \phi .$

Proof. Properties (i)$_\pm$, (ii)$_\pm$, and (iv) follow directly from definition (6.3). It further follows from the definition that $\cup\, E_+(t)$ is the set of all points free in the forward direction and this shows that $\cup\, E_+(t)$ is contained in Ω. According to Theorem 6.2 almost all points of Ω are free forward and hence (iii)$_+$ holds. The proof of (iii)$_-$ is similar.

We turn now to the Koopman representation of this flow; that is we denote by H the Hilbert space $L_2(\Omega, m)$ and define the operators $\{U(t)\}$ by

$$(6.4) \qquad\qquad [U(t)f](\omega) = f(\tau_{-t}(\omega)).$$

Since the flow is one-to-one, onto, and measure preserving, each operator $U(t)$ is a unitary map of H onto H; it is easy to show, in spite of the occasional discontinuity of the flow, that the $\{U(t)\}$ define a strongly continuous one-parameter group of operators.

Corresponding to the subsets $E_\pm$ we now define the following subspaces:

$$(6.5)_+ \qquad\qquad D_+ = [f\,;\,\operatorname{supp} f \subset E_+],$$

$$(6.5)_- \qquad\qquad D_- = [f\,;\,\operatorname{supp} f \subset E_-].$$

Theorem 6.4. *The subspaces D_- and D_+ satisfy the following properties:*

(i)$_-$ $U(t)D_- \subset D_-$ *for* $t \leq 0$,

(i)$_+$ $U(t)D_+ \subset D_+$ *for* $t \geq 0$,

(ii)$_-$ $\cap\, U(t)D_- = \{0\}$,

(ii)$_+$ $\cap\, U(t)D_+ = \{0\}$,

(iii)$_-$ $\cup\, U(t)D_-$ *is dense in* H,

(iii)$_+$ $\cup\, U(t)D_+$ *is dense in* H,

(iv) D_- *is orthogonal to* D_+.

Proof. Properties (i)$_\pm$, (ii)$_\pm$, and (iv) follow directly from the corresponding parts of Theorem 6.3. If (iii)$_+$ were not valid, then there would exist an f in H orthogonal to $\cup\, U(t)D_+$. We could infer from this that $U(t)f$ is orthogonal to D_+ for all t and this in turn requires, aside from a null set,† that

$$(6.6) \qquad\qquad \tau_t(\operatorname{supp} f) \cap E_+ = \phi$$

for all t. According to property (iii) of Theorem 6.3 this can happen only

---

† Since $E_+(t) \subset E_+$ for $t \geq 0$, the relation (6.6) holds for all t if it holds for all integral values of t and this requires throwing out only a denumerable number of null sets from the support of f.

if $m(\operatorname{supp} f) = 0$; that is, if f is the zero vector. Property (iii)$_-$ is proved similarly.

Theorem 6.4 asserts that D_- and D_+ are *incoming* and *outgoing* subspaces for the group $\{U(t)\}$ defined by the relation (6.4). According to the general theory, associated with each incoming [outgoing] subspace there is a translation representation of $\{U(t)\}$ mapping H unitarily onto $L_2(-\infty, \infty; N)$, where N is an auxiliary Hilbert space, sending $D_-[D_+]$ onto $L_2(-\infty, 0; N)$ [respectively $L_2(0, \infty; N)$] and the action of $U(t)$ into translation to the right by t units.

All of this applies in particular to the unperturbed problem for which there is no obstacle. In this case the phase space is $\Omega_0 = R_n \times S_{n-1}$ and the corresponding flow is simply

$$\tau_t^0(\{x, \theta\}) = \{x + t\theta, \theta\} .$$

The unperturbed group of unitary operators defined on $H_0 = L_2(\Omega_0, m)$ is defined as

$$[U_0(t)f](\omega) = f(\tau_{-t}^0(\omega)) .$$

It is convenient to replace $E_\pm$ by

$$E_-^0 = [\{x, \theta\} ; x \cdot \theta \le 0] \quad \text{and} \quad E_+^0 = [\{x, \theta\} ; x \cdot \theta \ge 0] ;$$

and $D_\pm$ by

$$D_-^0 = [f ; \operatorname{supp} f \subset E_-^0] \quad \text{and} \quad D_+^0 = [f ; \operatorname{supp} f \subset E_+^0] .$$

Denote by M the tangent manifold of the sphere S_{n-1} and by N the space $L_2(M)$ with respect to the usual measure in M. To each function f in $L_2(\Omega_0)$ we assign a function k_0 in $L_2(-\infty, \infty; N)$ as follows: Decompose x as

$$x = s\theta + \xi, \quad \xi \text{ orthogonal to } \theta ,$$

and set

(6.7) $$k_0(s) = f(\{s\theta + \xi, \theta\}) .$$

It is not hard to verify that the mapping:

$$f \rightarrow k_0$$

given by (6.7) is a unitary translation representation for $\{U_0(t)\}$ which is both incoming and outgoing.

It is clear for f in D_+ and for $t \geq 0$ that $U(t)f = U_0(t)f$ and hence for such f the action of the wave operator [see (1.2)] is given by $W_+ f = f$. Moreover, for any f with bounded spatial support, $U_0(t)f$ will lie in D_+ for t sufficiently large, say $t \geq T$, and hence for such an f, $W_+ f = U(-T)U_0(T)f$. Consequently, W_+ exists and is isometric on a dense subset of H_0 and can therefore be defined by continuity on all of H_0; the resulting W_+ is again an isometry. Finally the wave operator W_+ is an intertwining operator [see (1.4)]; that is

$$U(t)W_+ = W_+ U_0(t) .$$

It follows from this together with property (iii)$_+$ of Theorem 6.4 that the range of W_+ is all of H. A similar argument applies to W_- and hence the scattering operator S given by (1.5) is well defined and unitary.

As explained in Section 2, the perturbed incoming and outgoing translation representers k_- and k_+ of any f in H can be expressed in terms of the unperturbed representation in H_0 and the wave operators: For any f in H let $k_0(s)$ be the unperturbed translation representer of $W_+^{-1}f$; then

$$k_+(s) = k_0(s + \rho) ;$$

similarly, if k_0 is the unperturbed translation representer of $W_-^{-1}f$ then

$$k_-(s) = k_0(s - \rho) .$$

Now for f in D_+, $W_+ f = f$; and hence in this case we can write, using the explicit expression (6.7) for the unperturbed representation, that

$$(6.8)_+ \qquad k_+(s; \xi, \theta) = f(\{\xi + (s + \rho)\theta, \theta\}) \qquad \text{for } f \text{ in } D_+ ;$$

similarly,

$$(6.8)_- \qquad k_-(s; \xi, \theta) = f(\{\xi + (s - \rho)\theta, \theta\}) \qquad \text{for } f \text{ in } D_- ,$$

where in both cases $\theta \cdot \xi = 0$.

Next, for an f such that $U(T)f$ belongs to D_+ for some T, the outgoing translation representer of $U(T)f$ can be determined from $(6.8)_+$; that of f itself is obtained by translating backward by an amount T. This recipe can be put into the following more compact form:

$$(6.9)_+ \qquad k_+(s; \xi, \theta) = f(p_+)$$

where the point $p_+ = p(s; \xi, \theta)$ is

$(6.10)_+$ $$p_+ = \tau_s(\{\xi + \rho\theta, \theta\}) .$$

Similarly,

$(6.9)_-$ $$k_-(s; \xi, \theta) = f(p_-)$$

where

$(6.10)_-$ $$p_- = \tau_s(\{\xi - \rho\theta, \theta\}) .$$

Since the set of f of the above kind is dense in H by property (iii) of Theorem 6.4, it follows by continuity that $(6.9)_\pm$ holds for all f in H.

As described in (6.10) the points p_+ and p_- are functions of s, ξ, θ; these relations can be inverted to give s_+, ξ_+, and θ_+ as functions of p, and similarly s_-, ξ_-, and θ_- can be obtained as functions of p. There is thus a well defined one-to-one correspondence between the triplets $\{s_-, \xi_-, \theta_-\}$ and $\{s_+, \xi_+, \theta_+\}$ corresponding to the same point p; and we can regard $\{s_-, \xi_-, \theta_-\}$ as functions of $\{s_+, \xi_+, \theta_+\}$. It is important to note that in this correspondence the difference $s_- - s_+$ is determined as a function of ξ_+ and θ_+ alone:

(6.11) $$s_-(s_+, \xi_+, \theta_+) - s_+ \equiv l(\xi_+, \theta_+) .$$

This is obvious since for any value of t the triplets $\{s_- + t, \xi_-, \theta_-\}$ and $\{s_+ + t, \xi_+, \theta_+\}$ both correspond to the point $p_t = \tau_t(p)$.

Comparing formulas $(6.9)_+$ and $(6.9)_-$ we conclude that

$$k_+(s_+; \xi_+, \theta_+) = k_-(s_+ + l; \xi_-, \theta_-) ;$$

we rewrite this by dropping the subscript $(+)$:

(6.12) $$k_+(s; \xi, \theta) = k_-(s + l; \xi_-, \theta_-) ,$$

where l, ξ_-, and θ_- are functions of ξ and θ.

These functions l, ξ_-, and θ_- give the following interesting geometric description of the scattering process: A particle which is originally just about to leave E_- from the point $\xi_- - \rho\theta_-$, traveling in the θ_--direction, will after bouncing around for a time l enter E_+ at the point $\xi + \rho\theta$ traveling in the direction θ. The quantity l is called the *sojourn time*.

We recall that the scattering operator S relates incoming and outgoing translation representers of a given f in H as [cf. relation (2.1)]:

$$S: k_- \to k_+ ;$$

and the scattering matrix relates the corresponding spectral representers. Taking the Fourier transform of (6.12) we obtain

(6.13) $\hat{k}_+(\sigma\,;\xi,\theta) \,=\, \exp\left[-il(\xi,\theta)\sigma\right]\hat{k}_-(\sigma\,;\xi_-,\theta_-)\,.$

This shows that the action of the scattering matrix $S(\sigma)$ is a point transformation in M followed by multiplication by $\exp\,(-il\sigma)$.

We define the possibly infinite-valued domain functional $l(G)$ as follows:

(6.14) $l(G) \,=\, \mathrm{supp}\ l(\xi,\theta)\,.$

From the explicit description (6.13) of $S(\sigma)$ we deduce

Theorem 6.5.   *If $l(G)$ is finite then $S(\sigma)$ can be extended as an analytic function to the whole complex σ-plane. If $l(G) \,=\, \infty$ then $S(\sigma)$ is analytic in the lower half-plane but the real axis is its natural boundary.*

We close this section with a brief discussion of the semigroup of operators

$$Z(t) \,=\, P_+U(t)P_-$$

associated with the group $\{U(t)\}$, D_+ ₐ and D_- as in Section 3; here P_+ and P_- are as usual projections onto the orthogonal complements of D_+ and D_-, respectively. It is easy to show that the domain $K \,=\, H \ominus (D_+ \oplus D_-)$ of the semigroup is $L_2(\Omega_\rho)$ where Ω_ρ is the set

(6.15) $\Omega_\rho \,=\, [\{x,\theta\}\,;\ \,|\,x\cdot\theta\,|\,\le\,\rho\}\cap\Omega,$

and that for every f in K

$$[Z(t)f](\omega) \,=\, f(\tau_{-t}(\omega))\cdot \qquad \text{for}\quad \omega\ \text{in}\ \Omega_\rho$$

(6.16) $=\, 0\cdot \qquad\qquad \text{otherwise}\,.$

Now a trajectory enters the set Ω_ρ at a point $\{x\ \ \theta\}$ where $x\cdot\theta \,=\, -\rho$ and leaves it at a point x', θ' where $x'\cdot\theta' \,=\, \rho$; according to the geometrical characterization given above the time spent in Ω_ρ is just l; hence the name sojourn time. Therefore, if the functional $l(G)$ defined in (6.14) as the longest sojourn time is finite, then it follows from (6.15) and (6.16) that $Z(l(G)) \,=\, 0$. In this case the spectrum of the infinitesimal generator B of $\{Z(t)\}$ is empty and hence according to the results of section three the scattering matrix can be continued analytically into the upper half-plane. This confirms what we found to be the case earlier by examining the scattering matrix directly. Conversely it can be shown that if $l(G) \,=\, \infty$, then the infinitesimal generator B has a pure continuous spectrum which fills out the whole left half-plane and likewise this fact can be used to show that the real axis is a natural boundary of holomorphism for the scattering matrix.

7. Notes and Remarks

Ja. G. Sinaï's proof of the translation representation theorem of Section 2 is based on the following theorem of von Neumann:

Theorem. *Let $\{U(t)\}$ be a one-parameter group of unitary operators, B a selfadjoint operator, and suppose that*

$$(7.1) \qquad U(t)BU(-t) = B - tI, \qquad for\ all\ real\ \ t.$$

Then H can be represented as $L_2(-\infty, \infty; N)$ in such a way that $U(t)$ corresponds to right translation by t units and B to multiplication by the independent variable.

The odd-looking equation (7.1) is Weyl's form of the quantum mechanical commutation relation; differentiation with respect to t gives the familiar Heisenberg form: $AB - BA = iI$, where iA denotes the infinitesimal generator of the group $\{U(t)\}$.

It is of interest to note that Sinaï's proof is reversible in the sense that the representation theorem can be used to deduce von Neumann's theorem. In fact, if $\{E_\lambda\}$ denotes the spectral resolution of the identity for B so that

$$B = \int \lambda\, dE_\lambda,$$

then

$$U(t)BU(-t) = \int \lambda\, d[U(t)E_\lambda U(-t)]$$

and

$$B - tI = \int (\lambda - t)\, dE_\lambda = \int \lambda\, dE_{\lambda+t}.$$

According to (7.1) the two operators on the left are equal and since the right sides give spectral resolutions for these two operators, it follows that the two spectral families of projections are the same:

(7.2) $$U(t)E_\lambda U(-t) = E_{\lambda+t}.$$

Since $\{E_t\}$ is a spectral family of projections, it has the familiar properties:

   (i)   The range of E_t is an increasing function of t;
   (ii) As t tends to $-\infty$ the range of E_t shrinks to the null vector;
   (iii) As t tends to $+\infty$ the range of E_t tends to H.

Let D denote the range of E_0 ; the relation (7.2) shows that $D(t) = U(t)D$ is the range of E_t . Thus the above three properties are just the properties required of an incoming subspace in the translation representation theorem; in the resulting representation $U(t)$ acts as a right shift of t units and D corresponds to $L_2(-\infty, 0; N)$. Therefore, E_0 is multiplication by the characteristic function of the negative axis and by (7.2) E_λ is multiplication by the characteristic function of $(-\infty, \lambda)$. If we substitute this into the spectral resolution of B we see that B is multiplication by the independent variable, as asserted.

CHAPTER II

Representation Theory and
the Scattering Operator

We shall be concerned with obtaining certain representation theorems for a group of unitary operators $\{U(t); \ -\infty < t < \infty\}$ acting on a Hilbert space H for which there is an *outgoing* subspace. A closed subspace D_+ is called outgoing if it has the following properties:

(i) $U(t)D_+ \subset D_+$ for all $t > 0$,

(ii) $\cap \ U(t)D_+ = \{0\}$,

(iii) $\overline{\cup \ U(t)D_+} = H$.

A prototype of the above situation is obtained by setting H equal to $L_2(-\infty, \infty; N)$, the space of square integrable functions on the real numbers whose values lie in some auxiliary Hilbert space† N, where $U(t)$ corresponds to right translation by an amount t, and D_+ is $L_2(0, \infty; N)$. We shall show that, conversely, whenever there is an outgoing subspace for the group $U(t)$ there is always a representation for H and $U(t)$ of this kind. We shall call this an *outgoing translation representation*.

The discrete analog of this problem is easily solved and since a simple transformation allows one to obtain the continuous parameter representation in terms of the discrete analog representation, we begin with the discrete case.

1.   The Discrete Case

Instead of a one-parameter group of unitary operators we are concerned in the discrete case with a single unitary operator V and all of its powers.

---

† The scalar product in N will be denoted by $(n, m)_N$ and the norm by $|n|_N$.

With respect to V a closed subspace D_+ will be called outgoing if

(i) $V D_+ \subset D_+$,   (ii) $\cap\, V^k D_+ = \{0\}$,   (iii) $\overline{\cup\, V^k D_+} = H$.

A prototype of this situation is when H is $l_2(-\infty,\, \infty;\, N)$, the space of sequences $\{y_k;\, -\infty < k < \infty\}$ whose values lie in an auxiliary Hilbert space N and for which $\sum |\, y_k \,|_N{}^2 < \infty$, V corresponds to a right shift, and D_+ is $l_2(0,\, \infty;\, N)$.

Theorem 1.1. *If D is outgoing with respect to the unitary operator V, then H can be represented isometrically as $l_2(-\infty,\, \infty;\, N)$ for some auxiliary Hilbert space N so that V goes into the right shift operator and D maps onto $l_2(0,\, \infty;\, N)$. This representation is unique up to an isomorphism of N.*

Proof. We take N to be the orthogonal complement of VD in D, in symbols

(1.1) $$N = D \ominus VD.$$

We shall prove that

(1.2) $$D = \sum_{k \geq 0} \oplus\, V^k N,$$

and

(1.3) $$H = \sum \oplus\, V^k N.$$

The operator V being unitary, it is clear that

(1.4) $$V^k N = V^k D \ominus V^{k+1} D.$$

Moreover, it follows from property (i) that $V^{k+1} D \subset V^k D$ for all k and hence that the $V^k N$ are mutually orthogonal and that D contains

$$M \equiv \sum_{k \geq 0} \oplus\, V^k N.$$

Now, if M is a proper subspace of D there will exist a nonzero $x \in D \ominus M$. Since x is orthogonal to N it must lie in VD, since it is also orthogonal to VN it must lie in $V^2 D$. Continuing this line of reasoning we see that $x \in \cap\, V^k D$, which is contrary to property (ii). This proves (1.2) and it follows from (1.2) that

$$V^k D = \sum_{j \geq k} \oplus\, V^j N \subset \sum \oplus\, V^j N$$

and this implies (1.3) since by property (iii) the $\cup\, V^k D$ is dense in H.

In order to define the desired isomorphism, we note that each $x \in H$ has a unique decomposition by (1.3) of the form

$$x = \sum \oplus V^k y_k$$

where each y_k belongs to N. Further,

$$| x |_H^2 = \sum | V^k y_k |_H^2 = \sum | y_k |_N^2$$

and each sequence of y_k's in N with $\sum | y_k |_N^2 < \infty$ defines an $x \in H$ in this way. Consequently, the mapping

$$x \longrightarrow \{y_k\}$$

is an isometry of H onto $l_2(-\infty, \infty; N)$ under which D maps onto $l_2(0, \infty; N)$. Further, it is clear that

$$Vx = \sum \oplus V^{k+1} y_k \longrightarrow \{y_{k-1}\},$$

so that V corresponds under this mapping to the right shift operator. Observe that the construction (1.1) of N and the decompositions (1.2) and (1.3) of D and H are canonical. Using this one can easily prove that the representation is unique up to an isomorphism of N.

Next we employ the Fourier transformation to obtain the spectral representation from the translation representation.

Corollary 1.1. *If D is outgoing with respect to the unitary operator V, then H can be represented isometrically as $L_2(0, 2\pi; N)$ for some auxiliary Hilbert space N so that V goes into multiplication by $\exp(i\theta)$ and D is mapped onto $H_2(N)$, the Hardy class of functions $f(\theta)$ whose kth Fourier coefficients vanish for all negative k's. This representation is unique up to an isomorphism of N.*

Remark. We shall call this representation an *outgoing spectral representation* for the operator V.

Proof. The mapping

$$\{y_k\} \in l_2(-\infty, \infty; N) \longrightarrow f(\theta) \equiv \sum y_k e^{ik\theta} \in L_2(0, 2\pi; N)$$

with

$$| f |_{L_2}^2 = \frac{1}{2\pi} \int | f(\theta) |_N^2 d\theta = \sum | y_k |_N^2 = | \{y_k\} |_{l_2}^2$$

clearly defines an isomorphism of $l_2(-\infty, \infty; N)$ onto $L_2(0, 2\pi; N)$ taking the right shift operator into multiplication by exp $(i\theta)$ and $l_2(0, \infty; N)$ onto $H_2(N)$. Combining this with the result of Theorem 1.1 we obtain the desired outgoing spectral representation. The uniqueness follows by taking the inverse map:

$$f(\theta) \in L_2(0, 2\pi; N) \to \{y_k = \frac{1}{2\pi} \int f(\theta) e^{-ik\theta} \, d\theta\} \in l_2(-\infty, \infty; N)$$

and applying the corresponding result for the outgoing translation representation.

An *incoming subspace* D_- is defined similarly with property (i) replaced by (i)$_-$ $V^{-1}D_- \subset D_-$, while properties (ii) and (iii) remain as stated. We note that if $D_-[D_+]$ is incoming [outgoing], then the orthogonal complement of $D_-[D_+]$ is outgoing [incoming]. An analogous translation representation holds; in this case we arrange the mapping so that D_- maps onto $l_2(-\infty, -1; N_-)$. In the corresponding spectral representation D_- maps onto the conjugate Hardy class $\bar{H}_2(N_-)$ which we take to be those functions $f \in L_2(0, 2\pi; N_-)$ whose kth Fourier coefficients vanish for all nonnegative k's. As the next theorem shows, the auxiliary Hilbert spaces N and N_- obtained in the outgoing and incoming representations, respectively, are unitarily equivalent and shall henceforth be identified. Actually, the theorem shows that such an equivalence holds under even more general conditions.

Theorem 1.2. *If, for a given unitary operator, there exist two translation representations, say $l_2(-\infty, \infty; N)$ and $l_2(-\infty, \infty; N')$, then N and N' are unitarily equivalent.*

Remark. The assertion of this theorem is a direct consequence of multiplicity theory. However, the special form of the given representations permits the following simple proof.

Proof. It suffices to show that N and N' are of the same dimension. Suppose that N is of dimension $\aleph$ and N' of dimension $\aleph'$. Then $l_2(-\infty, \infty; N)$ and $l_2(-\infty, \infty; N')$ are clearly of dimension $\aleph_0 \cdot \aleph$ and $\aleph_0 \cdot \aleph'$, respectively. Since these two spaces are assumed to be isomorphic and hence of the same dimension, we conclude that if $\aleph$ and $\aleph'$ are both infinite then they are necessarily equal. Next suppose that $\aleph$ is finite and less than $\aleph'$. At this point it is convenient to work with the spectral representations, $L_2(0, 2\pi; N)$ and $L_2(0, 2\pi; N')$, respectively, rather than the given translation representations. Choose an orthonormal base $\{n_j ;$ $j = 1, \cdots, \aleph\}$ for N. Then the functions

$$\{e^{ik\theta}n_j ; \quad \text{all} \quad k \quad \text{and} \quad j = 1, \cdots, \aleph\}$$

form a complete orthonormal basis for $L_2(0, 2\pi; N)$ and hence if

$$n_j \in L_2(0, 2\pi; N) \rightarrow n_j'(\theta) \in L_2(0, 2\pi; N'),$$

then the functions

(1.5) $$\{e^{ik\theta}n_j'(\theta); \quad \text{all} \quad k \quad \text{and} j = 1, \cdots, \aleph\}$$

form a complete orthonormal basis for $L_2(0, 2\pi; N')$, since both spaces are spectral representations of the same unitary operator. In particular,

$$\frac{1}{2\pi} \int e^{ik\theta}[n_{j'}(\theta), n_m'(\theta)]_{N'} \, d\theta = \delta_{0k} \, \delta_{jm}.$$

As a consequence $[n_j'(\theta), n_m'(\theta)]_{N'} = \delta_{jm}$ almost everywhere; thus, for almost all θ, the vectors $\{n_j'(\theta); \quad j = 1, \cdots, \aleph\}$ form an orthonormal subset of N'. Choose any $\aleph + 1$ linearly independent vectors $\{m_k; \quad k = 1, \cdots, \aleph + 1\}$ in N'. For no θ can the $\{n_j'(\theta)\}$ span the subspace spanned by the $\{m_k\}$. Hence, for some k

$$m_k - \sum_{j=1}^{\aleph} [m_k, n_j'(\theta)]_{N'} n_j'(\theta)$$

must be different from zero on a set of positive measure. Call this function $g(\theta)$. Then $g(\theta)$ is measurable and $|g(\theta)|_{N'} \leq |m_k|_{N'}$, so that g is a nontrivial element of $L_2(0, 2\pi; N')$. Moreover, $[g(\theta), e^{ik\theta}n_j'(\theta)]_{N'} = 0$ and a fortiori g is L_2 orthogonal to the complete orthonormal basis formed by (1.5). We have thus arrived at a contradiction, proving that $\aleph = \aleph'$

2. The Scattering Operator in the Discrete Case

Let D_+ and D_- be outgoing and incoming subspaces for the same unitary operator, and suppose in addition that D_+ and D_- are *orthogonal*. To each vector $f \in H$ there are associated two vectors k_- and k_+, the respective incoming and outgoing translation representatives of f. The operator

$$S: k_- \rightarrow k_+$$

mapping $l_2(-\infty, \infty; N)$ onto $l_2(-\infty, \infty; N)$ is called the *scattering operator*.

Lemma 2.1.   *The scattering operator has the following properties:*

   (i)   *S is unitary;*

   (ii)   *S commutes with translation;*

   (iii)   *S maps $l_2(-\infty, -1; N)$ into $l_2(-\infty, -1; N)$.*

Proof.   Properties (i) and (ii) follow directly from the fact that S is defined in terms of two different unitary translation representations of the same operator. To deduce property (iii) we note that $k_- \in l_2(-\infty, -1; N)$ means that $f \in D_-$. Since we have assumed that D_- is orthogonal to D_+, it follows that f is orthogonal to D_+. On the other hand, in the outgoing translation representation D_+ is mapped onto $l_2(0, \infty; N)$ and hence k_+ lies in the orthogonal complement of the image of D_+ which is $l_2(-\infty, -1; N)$.

Corollary 2.1.   *Define the operator S as FSF^{-1}, F being Fourier transformation, that is S takes the incoming spectral representation onto the outgoing spectral representation. Then S has the following properties:*

   (i)′   *S is unitary;*

   (ii)′   *S commutes with multiplication by bounded measurable complex-valued functions;*

   (iii)′   *S maps $\bar{H}_2(N)$ into $\bar{H}_2(N)$.*

Proof.   Each of these properties except perhaps (ii)′ is a restatement of the corresponding property in the lemma. A restatement of (ii) is that S commutes the multiplicative operators $\{\exp(ik\theta)\}$. Property (ii)′ now follows from the fact that any bounded measurable scalar can be approximated pointwise almost everywhere and boundedly by linear combinations of $\{\exp(ik\theta)\}$ and that convergence of this sort implies the strong convergence of the corresponding multiplicative operators.

A linear operator on $L_2(0, 2\pi; N)$ such as S which commutes with multiplication by bounded measurable scalars and which maps $\bar{H}_2(N)$ into itself is called a causality operator by the physicists. It follows as a special case of a theorem due to Foures and Segal [1] that any causality operator S can be realized as a multiplicative operator-valued function $S(w)$ mapping N into N for each w with $|w| \geq 1$ and which is analytic for $|w| > 1$.

Theorem 2.1.   *Assuming N to be separable, an operator S satisfying properties (i)′–(iii)′ can be realized as a multiplicative operator-valued func-*

tion $S(e^{i\theta})$ *on N into N having the following properties:*

(a) $S(e^{i\theta})$ *is the boundary value of an operator-valued function* $S(w)$ *analytic for $|w| > 1$ which converges strongly†along radial rays for almost all θ to* $S(e^{i\theta})$;

(b) $|S(w)| \leq 1$ *for all w of absolute value >1;*

(c) $S(e^{i\theta})$ *is unitary for almost all θ.*

For the sake of completeness we include the following

Proof. Let n be any element of N. Then exp $(-i\theta)n$ belongs to $\bar{H}_2(N)$ and so, by (iii)′ $S[\exp(-i\theta)n]$ also belongs to $\bar{H}_2(N)$. Since

$$S[\exp(-i\theta)n] = \exp(-i\theta)Sn$$

by (ii)′, it follows that

$$Sn \equiv \tilde{n}(e^{i\theta})$$

belongs to exp $(i\theta)\bar{H}_2(N)$; such functions $\tilde{n}(e^{i\theta})$ are boundary values of vector-valued analytic functions $\tilde{n}(w)$ which are regular for $|w| > 1$.

Next, we show that $\tilde{n}(w)$ is uniformly bounded in $|w| > 1$. To this end let n and m be any pair of elements in N. Then for any integer $k \neq 0$ it is clear that exp $(ik\theta)$ n and m are orthogonal elements of $L_2(0, 2\pi; N)$. According to (i)′ the operator S is an isometry and hence it follows that the images of exp $(ik\theta)$ n and m are also orthogonal:

$$\int_0^{2\pi} e^{ik\theta}(\tilde{n}(e^{i\theta}), \tilde{m}(e^{i\theta}))_N \, d\theta = 0 \qquad (k \neq 0).$$

This means that all but the zeroth Fourier coefficient of the scalar integrable function $(\tilde{n}(e^{i\theta}), \tilde{m}(e^{i\theta}))_N$ are zero; such a function is a constant. The value of this constant is $(2\pi)^{-1}$ times the integral of the function which in this case is simply the scalar product of $\tilde{n}$ and $\tilde{m}$, and this in turn equals the scalar product of n and m. Thus

(2.1) $$(\tilde{n}(e^{i\theta}), \tilde{m}(e^{i\theta}))_N = (n, m)_N \qquad \text{a.e.}$$

Setting $n = m$ we obtain in particular

(2.1)′ $$|\tilde{n}(e^{i\theta})|_N = |n|_N \qquad \text{a.e.}$$

† A sequence of operators $\{T_k\}$ on N is said to converge strongly to T if $T_k n \to Tn$ for all n in N.

Since n belongs to $\exp (i\theta)\bar{H}_2(N)$, it can be represented in terms of its boundary values by the Poisson formula:

$$\tilde{n}(w) = \int_0^{2\pi} P(w, e^{i\theta})\tilde{n}(e^{i\theta})\, d\theta.$$

Using (2.1)′ and the fact that the Poisson kernel P is positive and has total weight one, we obtain the estimate

(2.2) $\qquad | \tilde{n}(w) |_N \leq \int P(w, e^{i\theta}) | \tilde{n}(e^{i\theta}) |_N\, d\theta = | n |_N.$

We now define the operator-valued function $S(w)$ for $| w | > 1$ as follows:

$$S(w)n \equiv \tilde{n}(w)$$

and claim that $S(w)$ has the properties asserted in the statement of Theorem 2.1. It is clear that $S(w)$ is strongly analytic in $| w | > 1$ and the inequality (2.2) shows that (b) is valid. In order to prove (a) we choose a denumerable dense subset $\{n_j\}$ of N. For each j, $S(w)n_j$ is a bounded function and so has radial limits as w tends to $\exp (i\theta)$ except on a subset E_j of measure zero; the union E of the sets E_j also has measure zero. Since the sequence $\{n_j\}$ is dense and since $S(w)$ is uniformly bounded in norm, it follows that $S(w)n$ has radial limits for any $n \in N$ except on the set E. Thus $S(w)$ has strongly defined boundary values which are contraction operators except on E; this concludes the proof of (a). If we adjoin to E the null set on which (2.1) fails to hold for all possible choices of $n = n_j$ and $m = n_k$, then (2.1) implies that except on E, $S(e^{i\theta})$ is an isometry on $\{n_j\}$ and hence by continuity on N.

The operation of multiplication by $S(e^{i\theta})$ is a bounded operation mapping $L_2(0, 2\pi; N)$ into itself. We claim that it is equal to the given operator S. Clearly, by construction, the two operators coincide for constant functions. Since both operators commute with multiplication by $\exp (i\theta)$, it follows that they also coincide for all trigonometric polynomials. Since trigonometric polynomials are dense in $L_2(0, 2\pi; N)$ and since both operators are bounded, we conclude that they agree on all of $L_2(0, 2\pi; N)$.

It remains to prove that $S(e^{i\theta})$ is unitary for almost all θ. For this purpose we choose a complete orthonormal basis $\{n_j\}$ for N and set $\tilde{n}_j(e^{i\theta}) = S(e^{i\theta})n_j$. For each θ not in E we know that $S(e^{i\theta})$ is an isometry and hence that $\{\tilde{n}_j(e^{i\theta})\}$ is an orthonormal set spanning the range of $\mathrm{S}(e^{i\theta})$ in N. If

$$n_k = \sum_j (n_k, \tilde{n}_j(e^{i\theta}))_N \tilde{n}_j(e^{i\theta}) \qquad \text{a.e.}$$

for each k, then the range of $\mathrm{S}(e^{i\theta})$ is almost everywhere equal to N and $\mathrm{S}(e^{i\theta})$ will be unitary almost everywhere. Otherwise there is a k such that

$$g(e^{i\theta}) \equiv n_k - \sum_j (n_k, \tilde{n}_j(e^{i\theta}))_N \tilde{n}_j(e^{i\theta})$$

differs from zero on a set of positive measure. Since $g(e^{i\theta})$ is measurable and since $|g(e^{i\theta})|_N \leq |n_k|_N$, we see that g is a nontrivial element of $L_2(0, 2\pi; N)$. Moreover,

$$(g(e^{i\theta}), e^{ik\theta}\tilde{n}_j(e^{i\theta}))_N = 0 \qquad \text{a.e.}$$

for all k and j, and *a fortiori* g is L_2-orthogonal to the set $\{e^{ik\theta}\tilde{n}_j(e^{i\theta})\}$. However, this set is the image under the unitary map S of the complete orthonormal basis $\{e^{ik\theta}n_j\}$ and hence also spans $L_2(0, 2\pi; N)$. This being impossible, it follows that $\mathrm{S}(e^{i\theta})$ is unitary for almost all θ.

Corollary 2.2. *The operator* $\mathrm{S}(w)$ *of Theorem 2.1 is determined to within right multiplication by a unitary operator on N by the properties* (i)′–(iii)′ *and the subspace* $D = \mathrm{S}\bar{H}_2(N)$.

Proof. Because of property (ii)′ the operator S is completely determined by its action on $N = e^{i\theta}\bar{H}_2(N) \ominus \bar{H}_2(N)$. It maps N onto $N' = e^{i\theta}D \ominus D$, a subspace isomorphic with N. Thus given any two unitary maps S_1 and S_2 satisfying the conditions of the corollary, it is clear that there exists a unitary map τ of N onto N, namely $\tau = \mathrm{S}_2^{-1}\mathrm{S}_1$ restricted to N, such that $\mathrm{S}_1 n = \mathrm{S}_2\tau n$ for all n in N, from which it follows that $\mathrm{S}_1(w) = \mathrm{S}_2(w)\tau$.

We will also have occasion to refer to the following restricted version of Theorem 2.1.

Corollary 2.3. *Assuming N to be separable, an operator S satisfying only the properties* (i)′ *and* (ii)′ *can be realized as a multiplicative operator-valued function* $\mathrm{S}(e^{i\theta})$ *on N into N which is unitary for almost all θ.*

Proof. Choose a complete orthonormal basis $\{n_j\}$ for N and set

$$\mathrm{S}(e^{i\theta})n_j \equiv \mathrm{S}n_j = \tilde{n}_j(e^{i\theta}).$$

As shown in the proof of the theorem, for each θ not in some null set E_0 the vectors $\{\tilde{n}(e^{i\theta})\}$ form a complete orthonormal basis for N. Any vector n in N can be expressed as

$$n = \sum a_k n_k \qquad \text{where} \quad \sum |a_k|^2 < \infty,$$

and we define

$$\mathcal{S}(e^{i\theta})n \equiv \sum a_k n_k(e^{i\theta}).$$

As so defined $\mathcal{S}(e^{i\theta})$ is obviously unitary for each θ in E_0 and a strongly measurable operator-valued function of θ. Finally, we note that the operation of multiplication by $\mathcal{S}(e^{i\theta})$ defines an isometry on $L_2(0, 2\pi; N)$ and since $\mathcal{S}$ and $\mathcal{S}(e^{i\theta})$ both commute with multiplication by scalar functions and coincide on constant vector-valued functions, they coincide on all of $L_2(0, 2\pi; N)$.

3.   The Continuous Case

We turn now to the representation problem for a strongly continuous one-parameter group of unitary operators $\{U(t)\}$ which we reduce to the representation of a single unitary operator treated in the first part of this chapter. The reduction is accomplished by means of the Cayley transform of the infinitesimal generator A of the group $\{U(t)\}$. It will be recalled that

$$Ax = \lim_{t \to 0+} \frac{U(t) - I}{t} x$$

and the domain of A, namely $D(A)$, consists of the set of all vectors for which this limit exists.

The relation

(3.1) $$(Ax, x) + (x, Ax) = 0, \qquad x \in D_A,$$

is a simple consequence of the fact that $|U(t)x| = |x|$. The resolvent set for A contains the right half-plane and in particular the point $\lambda = 1$; hence the range of $(I - A)$, in symbols, $R(I - A)$, is all of H. On the other hand the generator of the semigroup $\{U(-t); t \geq 0\}$ is $-A$ so that $R(I + A) = H$ as well. As a consequence, the Cayley transform

$$V = (I + A)(I - A)^{-1}$$

is a mapping with domain and range equal to all of H. If $x \in D(V)$ then there is a $y \in D(A)$ such that

$$(3.2) \qquad x = y - Ay \qquad \text{and} \qquad Vx = y + Ay.$$

Making use of (3.1) we see that

$$| x |^2 = | y |^2 + | Ay |^2 = | Vx |^2$$

and since $D(V) = H = R(V)$ it follows that V is a unitary operator. It is clear that the generator A can easily be recovered from V by inverting the Eqs. (3.2):

$$(3.3) \qquad y = 2^{-1}(Vx + x) \qquad \text{and} \qquad Ay = 2^{-1}(Vx - x).$$

We note that the Cayley transform of $-A$ is simply V^{-1}.

Lemma 3.1. *If $U(t)D \subset D$ for all $t > 0$, then $VD \subset D$, and conversely.*

Proof. Setting $R(\lambda, A) = (\lambda I - A)^{-1}$ we see that

$$(3.4) \qquad V = R(1, A) + AR(1, A) = 2R(1, A) - I,$$

and making use of the Laplace transform representation of $R(1, A)$ we have

$$(3.5) \qquad Vx = 2 \int_0^\infty e^{-t}U(t)x\, dt - x.$$

If for $x \in D$ we assume that $U(t)x \in D$ for all $t > 0$, then it follows from (3.5) that $Vx \in D$.

To prove the converse we first show that $R(\lambda, A)D \subset D$ for all $\lambda > 0$. Now the resolvent is analytic on the resolvent set and can be expanded in a power series:

$$(3.6) \qquad R(\lambda, A) = \sum_{n=0}^\infty (\lambda_0 - \lambda)^n [R(\lambda_0, A)]^{n+1}$$

valid for $| \lambda_0 - \lambda | | R(\lambda_0, A) | < 1$. For $\lambda_0 > 0$, $| R(\lambda_0, A) | \leq 1/\lambda_0$ so that the above series holds for $| \lambda - \lambda_0 | < \lambda_0$. It follows from this expansion that $R(\lambda_0, A)D \subset D$ implies $R(\lambda, A)D \subset D$ for all $| \lambda - \lambda_0 | < \lambda_0$. Assuming $VD \subset D$, one infers from (3.4) that $R(1, A)D \subset D$ and hence by a stepwise process using (3.6) that $R(\lambda, A)D \subset D$ for all $\lambda > 0$. Hence, for x in D and y in the orthogonal complement of D

$$0 = (R(\lambda, A)x, y) = \int_0^\infty e^{-\lambda t}(U(t)x, y)\, dt$$

for all $\lambda > 0$. By the Laplace transform uniqueness theorem $(U(t)x, y) = 0$ and hence $U(t)D \subset D$ for all $t > 0$. This completes the proof of the lemma.

Corollary 3.1. $U(t)M \subset M$ *for all real t if and only if $V^k M \subset M$ for all integers k. In either case $U(t)M = M = V^k M$ for all t and k.*

Proof. Applying the lemma to both the semigroup $\{U(t);\ t \geq 0\}$ and the semigroup $\{U(-t);\ t \geq 0\}$, we see that $U(t)M \subset M$ for all $t > 0$ (or $t < 0$) if and only if $VM \subset M$ (or $V^{-1}M \subset M$). The second assertion follows directly from the group property of $\{U(t)\}$ and $\{V^k\}$.

Lemma 3.2. *A closed subspace D is outgoing (or incoming) for $\{U(t)\}$ if and only if it is outgoing (or incoming) for V.*

Proof. The equivalence of property (i) in the two cases is a restatement of Lemma 3.1. In order to prove the equivalence of property (ii) in the two cases we consider the subspaces

$$P = \cap\, U(t)D \qquad \text{and} \qquad P' = \cap\, V^k D\,.$$

It is readily verified that $U(t)P = P$ for all t and that $V^k P' = P'$ for all k. According to the corollary to Lemma 3.1, $U(t)P' = P'$ for all t and $V^k P = P$ for all k. Finally, since P and P' are each subsets of D, we have

$$P = \cap\, V^k P \subset \cap\, V^k D = P' = \cap\, U(t)P' \subset \cap\, U(t)D = P\,.$$

Thus, $P = P'$, which implies the equivalence of property (ii) in the two cases. Property (iii) is treated in a similar fashion. In this case we set

$$M = \overline{\cup\, U(t)D} \qquad \text{and} \qquad M' = \overline{\cup\, V^k D}\,.$$

Since M and M' each contain D, we end up with the relation

$$M = \overline{\cup\, V^k M} \supset \overline{\cup\, V^k D} = M' = \overline{\cup\, U(t)M'} \supset \overline{\cup\, U(t)D} = M\,,$$

from which we infer the equivalence of the two statements of property (iii).

The above lemmas allow us to use the discrete representation theory already developed. In order to get from the $L_2(0, 2\pi; N)$ spectral representation of V to the appropriate $L_2(-\infty, \infty; N)$ spectral representation of $\{U(t)\}$ we make use of a fractional linear transformation taking the unit disk into the upper half-plane:

$$z = i\frac{1-w}{1+w} \quad \left(\text{with inverse } w = \frac{1+iz}{1-iz}\right).$$

In particular,

(3.7)
$$e^{i\theta} \to \sigma = i\frac{1-e^{i\theta}}{1+e^{i\theta}}$$

and

$$d\theta = 2(1+\sigma^2)^{-1}\,d\sigma.$$

We now define the following mapping:

(3.8) $g(e^{i\theta}) \in L_2(0, 2\pi; N) \to f(\sigma)$

$$= \pi^{-1/2}(1-i\sigma)^{-1}g\left(\frac{1+i\sigma}{1-i\sigma}\right) \in L_2(-\infty, \infty; N).$$

It is clear that this is an isometry, in fact

$$\frac{1}{2\pi}\int_0^{2\pi}|g(e^{i\theta})|_N^2\,d\theta = \int_{-\infty}^{\infty}|f(\sigma)|_N^2\,d\sigma.$$

Moreover, the mapping is obviously onto since the inverse mapping is well defined. We note that

$$e^{i\theta}g(e^{i\theta}) \to \frac{1+i\sigma}{1-i\sigma}f(\sigma).$$

Denote by $A_+(N)$ and $A_-(N)$, or simply A_+ and A_-, the images of the subspaces $H_2(N)$ and $\bar{H}_2(N)$, respectively, under the mapping (3.8). Since $H_2(N)$ and $\bar{H}_2(N)$ together span $L_2(0, 2\pi; N)$, the subspaces A_+ and A_- span $L_2(-\infty, \infty; N)$. Every function in $H_2(N)$ [or $\bar{H}_2(N)$] is the boundary value in the L_2 sense of an analytic function inside [outside] the unit disk whose square integral on concentric circles is uniformly bounded. It is easy to deduce from this and the explicit form of the mapping (3.8) that every function in A_+ [or A_-] is the boundary value in the L_2 sense of an analytic function in the upper [lower] half-plane whose square integral along the lines Im $z = $ const $> 0[<0]$ is uniformly bounded. According to the Paley–Wiener theorem, each such function is the Fourier transform of a square integrable function on the positive [negative] real axis, and conversely:

$$A_+ = FL_2(0, \infty; N), \qquad A_- = FL_2(-\infty, 0; N).$$

Thus we have proved

Lemma 3.3. *The relation* (3.8) *is an isomorphism of* $L_2(0, 2\pi; N)$ *onto* $L_2(-\infty, \infty; N)$ *mapping* $H_2(N)$ *onto* A_+ *and the multiplicative operator* $\exp(i\theta)$ *into the multiplicative operator* $(1 + i\sigma)(1 - i\sigma)^{-1}$.

We come now to the main result of this section.

Theorem 3.1. *If* D *is outgoing with respect to* $\{U(t)\}$, *then* H *can be represented isometrically as* $L_2(-\infty, \infty; N)$ *for some auxiliary Hilbert space* N *so that* $U(t)$ *goes into multiplication by* $\exp(i\sigma t)$ *and* D *is mapped onto* A_+. *This representation is unique up to an isomorphism on* N.

Remark. A representation of this kind is called an *outgoing spectral representation* for the group $\{U(t)\}$.

Proof. In order to prove this theorem we need only put together the construction of the previous lemma and the corresponding theorem for the discrete case. According to Lemma 3.2, D is outgoing for the Cayley transform V of the generator of $\{U(t)\}$. Applying Theorem 1.1 and its corollary we obtain the spectral representation for V in $L_2(0, 2\pi; N)$. In this spectral representation, D is mapped onto $H_2(N)$ and V goes into multiplication by $\exp(i\theta)$. Lemma 3.3 furnishes us with a transformation which maps $L_2(0, 2\pi; N)$ onto $L_2(-\infty, \infty; N)$, $H_2(N)$ onto A_+, and multiplication by $\exp(i\theta)$ into multiplication by $(1 + i\sigma)(1 - i\sigma)^{-1}$. It is readily verified that $(1 + i\sigma)(1 - i\sigma)^{-1}$ is the Cayley transform of the operation of multiplication by σ, which is the generator of the group of unitary multiplicative operators $\{\exp(i\sigma t)\}$. We recall that the Cayley transform uniquely determines the generator and that the generator uniquely determines the semigroup or group which it generates. Thus knowing as we do that $\{U(t)\}$ is mapped into a group of unitary operators the Cayley transform of whose generator is $(1 + i\sigma)(1 - i\sigma)^{-1}$, we can assert that $U(t)$ itself is mapped into $\exp(i\sigma t)$ in this representation.

If for a given outgoing subspace D there are two distinct outgoing spectral representations for the group $\{U(t)\}$, then we can retrace the above steps and obtain two distinct outgoing translation representations for V. According to Theorem 1.1 these translation representations of V can differ only by an isomorphism of N, and hence returning to the given spectral representations of $\{U(t)\}$, we see that they differ only by an isomorphism of N.

We now apply the inverse Fourier transform. According to the Paley–Wiener theorem quoted earlier, A_+ is mapped onto $L_2(0, \infty; N)$; thus we deduce

Corollary 3.2. *If D is outgoing with respect to the group of unitary operators $\{U(t)\}$, then H can be represented isometrically as $L_2(-\infty, \infty; N)$ for some auxiliary Hilbert space N so that $U(t)$ goes into translation to the right by t units, and D is mapped onto $L_2(0, \infty; N)$. This representation is unique up to an isomorphism of N.*

Remark. A representation of this kind is called an *outgoing translation representation* for the group $\{U(t)\}$.

Analogous representation theorems hold for an incoming subspace D_-. In the incoming spectral representation D_- is mapped onto A_- and in the incoming translation representation D_- is mapped onto $L_2(-\infty, 0; N)$. As a special case of the next theorem, we see that the auxiliary Hilbert spaces N and N_- obtained in the outgoing and incoming representations, respectively, are unitarily equivalent and shall henceforth be identified.

Theorem 3.2. *If, for a given unitary group $\{U(t)\}$, there exist two translation (or spectral) representations, say $L_2(-\infty, \infty; N)$ and $L_2(-\infty, \infty; N')$, then N and N' are unitarily equivalent.*

Proof. This follows as above from Theorem 1.2. We simply map these spaces onto $L_2(0, 2\pi; N)$ and $L_2(0, 2\pi; N')$, respectively, and obtain two spectral representations for the Cayley transform V of the generator of $\{U(t)\}$.

4. The Scattering Operator in the Continuous Case

Just as in the discrete case, we begin with orthogonal outgoing and incoming subspaces D_+ and D_- for a given group of unitary operators $\{U(t)\}$. Again to each vector $f \in H$ there are associated two vectors k_- and k_+, the respective incoming and outgoing translation representers of f. The mapping

$$S: k_- \rightarrow k_+$$

is called the *scattering operator*.

Lemma 4.1.   *The scattering operator has the following properties:*

(i) S *is unitary;*
(ii) S *commutes with translation;*
(iii) S *maps* $L_2(-\infty, 0; N)$ *into itself.*

The proof paraphrases that of Lemma 2.1. In fact, if we retrace our steps back to the translation representation for the Cayley transform V of the generator of $\{U(t)\}$, we see that S is a representation of the scattering operator previously defined in Section 2 for V. If we proceed to the incoming and outgoing spectral representations for V we obtain the operator which we now denote by S' described in Corollary 2.1. This operator can be realized as a multiplicative operator-valued function $S'(e^{i\theta})$ having the properties listed in Theorem 2.1 and Corollary 2.2. Next, making use of the mapping from $L_2(0, 2\pi; N)$ to $L_2(-\infty, \infty; N)$ defined in Lemma 3.3, we obtain a representation S for S from the incoming to the outgoing spectral representations of $\{U(t)\}$ having the following properties:

(i)″ S *is unitary;*
(ii)″ S *commutes with multiplication by bounded measurable complex-valued functions;*
(iii)″ S *maps* A_- *into* A_- .

We now interpret S as an operator on $L_2(-\infty, \infty; N)$ to itself. Setting

$$S(z) = S'\left(\frac{1 + iz}{1 - iz}\right),$$

Theorem 2.1 is transformed into:

Theorem 4.1.   *Assuming N to be separable, an operator S on $L_2(-\infty, \infty; N)$ satisfying the properties (i)″–(iii)″ can be realized as a multiplicative operator-valued function $S(\sigma)$ on N into N having the following properties:*

(a) $S(\sigma)$ *is the boundary value of an operator-valued function $S(z)$ analytic for Im $z < 0$ which converges strongly along the lines Re $z = \sigma$ to $S(\sigma)$ for almost all σ;*

(b) $|S(z)| \leq 1$ *for all z with* Im $z < 0$;

(c) $S(\sigma)$ *is unitary for almost all* σ.

Remark.   Actually, radial convergence in part (a) of Theorem 2.1 is transformed into convergence along a circle intersecting the real axis at right angles. However, because of analyticity and the inequality (b) we obtain from the usual estimate of the derivative the inequality

$$| S(\sigma_1 - i\eta) - S(\sigma_2 - i\eta) | \leq | \sigma_1 - \sigma_2)/\eta ,$$

and this together with convergence along circles implies convergence along the lines Re $z = \sigma$.

Corollaries 2.2 and 2.3 become:

Corollary 4.1.   *The operator* $S(z)$ *of Theorem 4.1 is determined to within right multiplication by a unitary operator on N by the properties* (i)''–(iii)'' *and the subspace $D = SA_-(N)$.*

Corollary 4.2.   *Assuming N to be separable, an operator S satisfying only the properties* (i)'' *and* (ii)'' *can be realized as a multiplicative operator-valued function $S(\sigma)$ on N into N which is unitary for almost all σ.*

Physicists customarily define the scattering operator in terms of a perturbed and an unperturbed group of unitary operators, $\{U(t)\}$ and $\{U_0(t)\}$, respectively. They begin with the wave operators

(4.1) $W_{\pm} = \text{strong lim } U(-t) U_0(t)$
                                      $t \to \pm\infty$

and then define the scattering operator S in terms of these as

(4.2) $S = W_+^{-1} W_- .$

We shall show that, given any group of unitary operators $\{U(t)\}$ and a pair of orthogonal incoming and outgoing subspaces D_- and D_+, we can define another group of unitary operators $\{U_0(t)\}$ so that if we regard $U(t)$ as a perturbation on $U_0(t)$ then the classical scattering operator (4.2) coincides with (actually differs by a trivial factor from) the scattering operator which we introduced in the first part of this section.

We define $U_0(t)$ to be right translation on $H_0 \equiv L_2(-\infty, \infty; N)$. To connect $U_0(t)$ with $U(t)$ we make the following identifications in the

Hilbert spaces on which they act: The subspace $L_2(-\infty, -1; N)$ of H_0 is identified with D_- in the incoming translation representation of $\{U(t)\}$ (after translating one unit to the left); the subspace $L_2(1, \infty; N)$ of H_0 is identified with D_+ in the outgoing translation representation of $\{U(t)\}$ (after translating one unit to the right); and the leftover portion $L_2(-1, 1; N)$ of H_0 is identified with $H \ominus (D_- \oplus D_+)$ of H in a unitary but otherwise arbitrary fashion.

We can now describe the wave operators: It is clear that when the support of $h(s) \in H_0$ is bounded from below, then for t sufficiently large, say $t \geq T$, $U_0(t)h$ has its support in $[1, \infty)$; and for $t \geq T$ the function $U(-t)U_0(t)h$ is *independent of* t and consequently the limit W_+h exists and is an isometry. Since the set of functions h with support bounded from below is dense in H_0, it follows that W_+ exists for all h in H_0. Similarly, if f in H has the property that for some T, $U(T)f$ belongs to D_+, then f will belong to the range of W_+. By assumption (iii) on D_+, the set of such elements is dense in H so that the range of W_+ is dense. The operator W_+ is therefore unitary.

The above analysis shows that if $f = W_+h$, then h is the outgoing translation representer of f, shifted to the right by one unit. Similarly, the other wave operator W_- provides the incoming translation representation shifted to the left by one unit. Hence the scattering operator defined in (4.2) does indeed relate incoming and outgoing translation representations of the functions in H. The extraneous shifts introduce a harmless exponential factor into the corresponding scattering matrix.

The perturbation considered in this example has the special feature that there exist substantial subspaces, D_- and D_+, over which $U_0(t)$ and $U(t)$ act in the same way for t negative and positive, respectively. In applications to potential scattering this corresponds to the fact that our method applies only to potentials with bounded support. Also the existence of incoming or outgoing subspaces implies, as the representation theorem shows, that the spectrum of the infinitesimal generator has uniform multiplicity over the whole real axis. This places another limitation on the kind of scattering problem which can be treated directly by our approach.

In the applications to potential and obstacle scattering to be described in Chapters IV and V, the unperturbed group $\{U_0(t)\}$, which was here constructed quite artificially, corresponds to wave propagation in free space. We shall in this case explicitly construct a translation representation for $\{U_0(t)\}$ which is both incoming and outgoing with respect to D_- and D_+. The unperturbed representer of f can be taken as the perturbed outgoing representer of f for all f in D_+, and as the perturbed incoming representer of f for all f in D_-. Since the translates of $D_+[D_-]$ are dense in H, this procedure leads to the construction of the incoming and outgoing representations for $\{U(t)\}$ without appeal to the general representation theorem.

The scope of our result can be extended by using Birman [2] and Kato's [2] principle of *invariance of the wave operator* which we present below in the context of our theory. Let $\{U(t)\}$ be a group of operators of the kind considered in this chapter and let $\{U_0(t)\}$ be the associated unperturbed group constructed above. Finally let ϕ be a real-valued function defined on $(-\infty, \infty)$ with the following properties: The real axis can be partitioned into a finite number of open intervals $\{I_k\}$ plus their end points such that ϕ is strictly monotonic and twice continuously differentiable in each I_k. In what follows we set $R_+[R_-]$ equal to the collection of intervals where ϕ is increasing [decreasing].

Next, let A and A_0 denote the infinitesimal generators of $\{U(t)\}$ and $\{U_0(t)\}$, respectively; let A' and A_0' denote the skew self-adjoint operators $i\phi(-iA)$ and $i\phi(-iA_0)$; and finally let the unitary groups generated by A' and A_0' be written as $\{U'(t)\}$ and $\{U_0'(t)\}$.

Theorem 4.2. *The wave operators $W_\pm'$ connecting the groups $\{U_0'(t)\}$ and $\{U'(t)\}$ exist and are unitary; furthermore $W_\pm'$ are "piecewise" equal to the wave operators $W_\pm$ connecting $\{U_0(t)\}$ and $\{U(t)\}$ in the following sense: Let f_+ be an element of H_0 whose spectrum with respect to A_0 lies in R_+; then*

$(4.3)_+$ $$W_\pm'f_+ = W_\pm f_+ .$$

Similarly,

$(4.3)_-$ $$W_\pm'f_- = W_\mp f_-$$

for all f_- whose spectrum lies in R_-.

Remark. Our proof is an adaptation of Kato's, somewhat simplified by imposing the assumption that ϕ be twice differentiable; this is not necessary for the validity of the result but is sufficient for the application we have in mind.

By definition (4.1)

(4.4) $$W_+'f = \lim_{t \to \infty} U'(-t) U_0'(t) f.$$

Since the operators on the right in (4.4) are uniformly bounded it suffices to prove the existence of the limit for a set of f which span H_0. The following set of functions f_+ and f_- clearly span H_0:

(i)   The representer $\tilde{f}_+(\sigma)[\tilde{f}_-(\sigma)]$ of $f_+[f_-]$ in the unperturbed (incoming–outgoing) spectral representation is a smooth function of σ;

(ii)   The support of $\tilde{f}_+[\tilde{f}_-]$ is contained in a compact subset of $R_+[R_-]$.

Lemma 4.2.   *For every function $\tilde{f}_+[\tilde{f}_-]$ satisfying the properties* (i) *and* (ii) *above the distance of $U_0'(t)f_+[U_0'(t)f_-]$ to $D_+[D_-]$ tends to zero as t tends to infinity.*

Proof. For convenience of notation we consider the $(+)$ case and perform the indicated orthogonal decomposition:

(4.5) $$U_0'(t)f_+ = g_1(t) + g_2(t) ,$$

$g_1(t)$ in D_+ and $g_2(t)$ orthogonal to D_+. We want to prove that $g_2(t)$ tends to zero as t becomes infinite.

In the unperturbed spectral representation the action of $U_0'(t)$ is multiplication by $\exp(i\phi(\sigma)t)$; whereas projection onto the orthogonal complement of D_+ corresponds to the restriction of the support of a function to the interval $(-\infty, 1)$ in the unperturbed *translation* representation. Since the spectral and translation representations are related by Fourier transformation we see that the translation representation of $g_2(t)$ is equal to

$$\int \exp(-i\sigma s) \exp[i\phi(\sigma)t]\tilde{f}_+(\sigma)\, d\sigma = \int \exp[i\phi(\sigma)t - i\sigma s]\tilde{f}_+(\sigma)\, d\sigma$$

for s in $(-\infty, 1)$ and to zero elsewhere. Now $\tilde{f}_+$ was assumed to be a once differentiable and ϕ a twice differentiable function of σ; we can therefore integrate by parts and obtain

(4.6) $$i \int \frac{\exp[i\phi(\sigma)t - i\sigma s]}{\dot{\phi}(\sigma)t - s} \dot{\tilde{f}}_+(\sigma)\, d\sigma$$
$$- it \int \frac{\exp[i\phi(\sigma)t - i\sigma s]}{(\dot{\phi}(\sigma)t - s)^2} \ddot{\phi}(\sigma)\tilde{f}_+(\sigma)\, d\sigma.$$

This formula is valid provided that the denominator does not vanish; but this is so since, on the one hand, the variable s is restricted to the interval $(-\infty, 1)$ and, on the other hand, the support of the function $\tilde{f}_+$ is restricted to a compact subset of R_+ over which $\dot{\phi}(\sigma)$ is bounded from below by some positive constant, say δ. Therefore,

$$|\dot{\phi}(\sigma)t - s| > |\delta t - s|;$$

and since $\tilde{f}_+$ and $\dot{\tilde{f}}_+$ were assumed to be integrable and $\ddot{\phi}$ was assumed to be bounded on the supp $\tilde{f}_+$, the integrals in (4.6) are bounded by const $(\delta t - s)^{-1}$. This is a bound for the value of the translation representation of $g_2(t)$ at the point s and the square integral with respect to s of this function on $(-\infty, 1)$ gives an upper bound for $|g_2(t)|^2$. Since

$$\int_{-\infty}^{1} \frac{1}{(\delta t - s)^2}\, ds = \frac{1}{\delta t - 1}$$

tends to zero as t becomes infinite, this completes the proof of Lemma 4.2.

We now substitute (4.5) into (4.4) and because we have already shown that $g_2(t)$ tends to zero we get

$$W_+' f_+ = \lim_{t \to \infty} U'(-t)[g_1(t) + g_2(t)] = \lim_{t \to \infty} U'(-t)g_1(t) .$$

Since g_1 belongs to D_+ its perturbed outgoing spectral representer is the same (except for a factor e^{-i} which we shall omit) as its unperturbed spectral representer. Now the unperturbed spectral representer of $g_1 + g_2 = U_0'(t)f_+$ is $\exp(i\phi(\sigma)t)\tilde{f}_+(\sigma)$; and consequently that of $g_1(t)$ is

$$\exp[i\phi(\sigma)t]\tilde{f}_+(\sigma)$$

plus a function which is negligible since its L_2-norm tends to zero as t tends to infinity. Since the action of $U'(-t)$ in the perturbed outgoing spectral representation is multiplication by $\exp(-i\phi(\sigma)t)$, it follows from the foregoing that the perturbed outgoing spectral representer of $U'(-t)g_1(t)$ is $\tilde{f}_+(\sigma)$ plus a negligible function. The limit of this at t tends to infinity is simply $\tilde{f}_+(\sigma)$; thus $W_+' f_+$ exists and its perturbed outgoing spectral representer is $\tilde{f}_+(\sigma)$.

On the other hand we have previously shown that the perturbed outgoing spectral representer of $W_+ f$ is equal to the unperturbed spectral representer of f; this proves one half of $(4.3)_+$. The corresponding half of $(4.3)_-$ can be proved similarly. In this case we are dealing with an element f_- whose unperturbed spectral representer has its support in R_-; and we split $U_0'(t)f_-$ into orthogonal parts $g_1(t)$ in D_- and $g_2(t)$ in the orthogonal complement of D_-. By Lemma 4.2

$$| g_2(t) |^2 = \int_{-1}^{\infty} \left| \int_{R_-} \exp(-i\sigma s) \exp[i\phi(\sigma)t] \tilde{f}_-(\sigma)\, d\sigma \right|^2 ds$$

tends to zero as t becomes infinite. Since $g_1(t)$ now lies in D_- its perturbed *incoming* spectral representer is the same (except for a factor e^i) as its unperturbed spectral representer and hence

$$W_+'f_- = \lim_{t\to\infty} U'(-t) U_0'(t)f_- = \lim_{t\to\infty} U'(-t)g_1(t)$$

will have $\tilde{f}_-$ as its perturbed incoming spectral representer; thus $W_+'f_- = W_-f_-$. The remaining halves of $(4.3)_\pm$ are proved in a similar fashion.

It is clear from the definition that the operators $W_\pm'$ are isometries. Hence in order to prove that they are unitary it suffices to show that their ranges are all of H. If $\{E(I)\}$ and $\{E_0(I)\}$ denote the resolutions of the identity for A and A_0, respectively, then it follows from the intertwining relation (1.4) of Chapter I that

$$E(I)W_\pm = W_\pm E_0(I).$$

In particular this holds for I replaced by R_+ or R_- and since $W_\pm$ are known to be unitary, we see that the range of $W_\pm E_0(R_\pm)$ coincides with $E(R_\pm)$. This concludes the proof of Theorem 4.2.

Corollary 4.3. *Let S' be the scattering operator for the ϕ-transformed pair; that is $S' = W_+'^{-1}W_-'$. Then $S'f_+ = Sf_+$ and $S'f_- = S^{-1}f_-$.*

In Chapter VI we shall use Theorem 4.2 to construct the scattering matrix for the Schrödinger wave equation from the scattering matrix for the related acoustic problem.

5. Notes and Remarks

The representation theorem derived in this chapter is not new; it was previously found by Sinaĭ in [1]. His brief and elegant proof is based on the famous von Neumann theorem on the uniqueness of the solutions to the Weyl form of the quantum mechanical commutation laws. It should be added that this is merely the starting point in Sinaï's paper; he goes on to consider unitary groups arising from so-called K-systems in probability theory, and shows for these that the auxiliary space N is always infinite dimensional. Our proof of the representation theorem is closely related to the methods introduced by Masani and Robertson [1] in their study of stationary stochastic processes.

CHAPTER III

A Semigroup of Operators Related to the Scattering Matrix

In the classical theory of scattering of sound waves and electromagnetic waves, the scattering operator relates the behavior of the solution to the unperturbed problem to that of the perturbed problem at large distances and large positive times for solutions which behave in the same way at large distances and large negative times in the two systems. In addition it was found that there is a relation between the behavior of solutions to the perturbed problem at fixed positions and large positive times and the analytic continuation of the scattering matrix into the upper half-plane. In particular a rapidly converging expansion, valid for fixed positions and large positive times, can be obtained in terms of the poles of the scattering matrix.

In this chapter we shed new light on this aspect of scattering theory with the aid of a family of operators $\{Z(t)\,; t \geq 0\}$ defined by

$$Z(t) \; = \; P_+ U(t) P_- \qquad (t \geq 0) \; ;$$

here P_+ is the orthogonal projection of H onto the orthogonal complement of D_+ and P_- is the orthogonal projection of H onto the orthogonal complement of D_-. The reason for looking at the operator $Z(t)$ is, in part, the following:

At the end of the last chapter we constructed a family of operators $\{U_0(t)\}$ relative to which $\{U(t)\}$ was a perturbation. The action of the two groups agreed on D_- for negative t and on D_+ for positive t. Thus we can regard D_- as consisting of signals unaffected in the past by the perturbation and D_+ as consisting of signals unaffected in the future by the

61

perturbation. In particular D_- contains some signals which will remain unaffected by the perturbation for a very long time and the purpose of the operator P_- is to remove these signals from consideration. After the action of $U(t)$, P_+ removes that part of the signal which will no longer be affected by the scatterer. Consequently the operators $\{Z(t)\}$ appears to be a reasonable tool for isolating the effects of the perturbation.

We shall show in this chapter that $\{Z(t)\}$ forms a semigroup of operators over the orthogonal complement of D_+ and D_-, and that all of the essential information pertaining to the relation between $\{U(t)\}$, D_+, and D_- can be recovered from $\{Z(t)\}$. In particular $\sqrt{-1}$ times the complex conjugate of the spectrum of the infinitesimal generator of $\{Z(t)\}$ coincides with the set of points in the lower half-plane at which the scattering matrix is not regular, and if $Z(T)$ is compact for some T, then the expansion of $Z(t)x$ in terms of the eigenvectors of the operators $Z(t)$ coincides with the previously mentioned expansion obtained from the poles of the scattering matrix.

In Chapter V we shall show for scattering by a bounded obstacle or by a positive potential of finite support that the spectrum of the infinitesimal generator of $\{Z(t)\}$ is discrete and it follows from this that the scattering matrix S is regular in the lower half-plane except for a discrete set of points where it has zeros. Since S is unitary on the real axis, it follows by the Schwarz reflection principle that S can be continued analytically into the upper half-plane so as to be meromorphic.

1.   The Related Semigroups

We begin by proving that the operators $Z(t)$ form a semigroup.

Theorem 1.1.   *The operators $\{Z(t)\,;\,t \geq 0\}$ annihilate D_+ and D_-, map the orthogonal complement K of $[D_+ \oplus D_-]$ into K, and form a strongly continuous semigroup of contraction operators on K. Furthermore, $Z(t)$ tends strongly to zero as $t \to \infty$: $\lim Z(t)x = 0$ for every x in K.*

Proof.   By definition $Z(t) = P_+U(t)P_-$ and since P_- annihilates D_- so does $Z(t)$. Moreover, since D_- is orthogonal to D_+, P_- acts like the identity on D_+; for t positive, $U(t)$ maps D_+ into itself, and consequently P_+ annihilates $U(t)D_+$; thus, $Z(t)$ also annihilates D_+.

Next we show that $Z(t)$ maps K into K. The last factor in $Z(t)$ being P_+ it follows that the range of $Z(t)$ is orthogonal to D_+. What remains to be shown is that if x is orthogonal to D_- then so is $Z(t)x$: Now $P_-x = x$ if x is orthogonal to D_- and since $U(-t)$ maps D_- into itself, its adjoint, namely $U(t)$, maps the orthogonal complement of D_- into itself. Thus $U(t)P_-x$ is orthogonal to D_- and finally since D_+ is orthogonal to D_-, the projection P_+ maps the orthogonal complement of D_- into itself so that $Z(t)x$ is also orthogonal to D_-.

In proving that $\{Z(t)\}$ forms a semigroup on K, we make use of the identity

$$P_+U(t)P_+ = P_+U(t) \qquad (t \geq 0) ,$$

which follows from the fact that $U(t)$ maps D_+ into itself. Note also that P_- acts like the identity on K so that $Z(t) = P_+U(t)$ on K. Thus,

$$Z(t)Z(s) = P_+U(t)P_+U(s) = P_+U(t)U(s) = Z(t + s)$$

on K. This proves the semigroup property. That $Z(t)$ is strongly continuous and a contraction follows directly from the corresponding properties of $U(t)$.

To show that $Z(t)$ tends strongly to zero we use the fact that $U(t)$-translates of D_+ are dense in H, that is property (iii) of an outgoing subspace. Thus given any x in K and $\epsilon > 0$ there exists an element y in D_+ and a positive number T such that $| x - U(-T)y | < \epsilon$. Since $P_+U(t)$ is a contraction and $P_+U(t)x = Z(t)x$ for x in K, we obtain

$$| Z(t)x - P_+U(t)U(-T)y | < \epsilon .$$

For $t > T$ we have, since y belongs to D_+, that $P_+U(t)U(-T)y = P_+U(t - T)y = 0$; so for $t > T$ $| Z(t)x | < \epsilon$. This completes the proof of Theorem 1.1.

Next consider the action of $Z(t)$ in the outgoing translation representation where $U(t)$ corresponds to right translation by t units and P_+ corresponds to restriction to the negative real axis. The action of $Z(t)$ is obviously translation followed by truncation. In this representation D_+ corresponds to $L_2(0, \infty ; N)$ so that its orthogonal complement is represented by $L_2(-\infty, 0; N)$. As we have seen in Section 4 of Chapter II, D_- is represented by $S[L_2(-\infty, 0; N)]$ and hence K is represented by

$$(1.1) \qquad K \leftrightarrow L_2(-\infty , 0 ; N) \ominus SL_2(-\infty , 0 ; N) .$$

As we have shown above, $Z(t)$ maps K into itself and hence the space representing K in the outgoing translation representation is invariant under right shift followed by truncation.

In the outgoing *spectral* representation K corresponds to a subspace of functions which we denote by R:

$$(1.2) \qquad\qquad R = A_- \ominus \mathcal{S}A_- ,$$

where A_- is as before the Fourier transform of $L_2(-\infty, 0; N)$.

Denote by A_+ the Fourier transform of $L_2(0, \infty; N)$. Since Fourier transform is unitary, A_+ is the orthogonal complement of A_-; similarly, since multiplication by $\mathcal{S}$ is unitary, $\mathcal{S}A_+$ is the orthogonal complement of $\mathcal{S}A_-$. So (1.2) can be written as

$$(1.2)' \qquad\qquad R = A_- \cap \mathcal{S}A_+ .$$

If we interchange the roles of D_+ and D_-, and at the same time change the direction of t, we again obtain a semigroup of operators, namely $\{P_-U(-t)P_+; t \geq 0\}$, to which Theorem 1.1 is applicable. This is simply the adjoint semigroup, that is

$$Z^*(t) = P_-U(-t)P_+ \qquad (t \geq 0) .$$

In the *incoming* translation representation K is represented by

$$(1.3) \qquad\qquad K \leftrightarrow L_2(0, \infty; N) \ominus S^{-1}L_2(0, \infty; N)$$

and $Z^*(t)$ acts as left translation followed by truncation. In the *incoming* *spectral* representation K corresponds to a subspace of A_+ which we denote by R^*. It follows from the definition and unitary character of $\mathcal{S}$ that

$$(1.4) \qquad\qquad R^* = A_+ \ominus \mathcal{S}^*A_+ = A_+ \cap \mathcal{S}^*A_- = \mathcal{S}^*R .$$

The following characterization of R follows immediately from $(1.2)'$.

Lemma 1.1. *The function $f(\sigma)$ belongs to R if and only if f belongs to A_- and $\mathcal{S}^*(\sigma)f(\sigma)$ belongs to A_+.*

The properties of $\mathcal{S}^*$ are entirely analogous to those of $\mathcal{S}$.

Lemma 1.2. $\mathcal{S}^*(\bar{z})$ *is an operator-valued function of z, analytic in the upper half-plane where it is of norm less than or equal to one. As z tends to σ from above on verticle segments, $\mathcal{S}^*(\bar{z})$ converges strongly to $\mathcal{S}^*(\sigma)$ for almost all real σ.*

Proof. Since $S(z)$ is analytic in the lower half-plane, the formula

$$(1.5) \qquad (S(z)n, m)_N = (n, S^*(z)m)_N$$

shows that $S^*(z)$ is weakly skew-analytic there. By a well-known principle (uniform boundedness) $S^*(z)$ is skew-analytic also in the uniform operator topology. Consequently $S^*(\bar{z})$ is analytic in the upper half-plane.

Adjoint operators have the same norm; so

$$| S^*(z) |_N = | S(z) |_N \leq 1 \qquad \text{for} \quad \text{Im } z \leq 0 .$$

According to Theorem 4.1 of Chapter II, if we except a null set of σ's, $S(z)$ tends strongly to $S(\sigma)$ as z tends to σ from below along vertical segments. It follows from (1.5) that $S^*(z)m$ tends weakly to $S^*(\sigma)m$ along such vertical segments. To show that this convergence is actually strong, we note that since $S^*(\sigma)$ is unitary for almost all σ, $| S^*(\sigma)m |_N = | m |_N$. On the other hand, as shown above, $| S^*(z)m |_N \leq | m |_N$. According to a well-known elementary principle of Hilbert space theory, if a sequence $\{u_k\}$ tends weakly to u and if

$$\lim \sup | u_k |_N \leq | u |_N ,$$

then the convergence is necessarily strong. Applying this principle to the above situation completes the proof.

The following further result about the behavior of $S(z)$ at the boundary of its domain of definition is needed:

Lemma 1.3. *Let σ_0 be a point on the real axis and suppose for all z in the lower half-plane and near enough to σ_0 that $S(z)$ is regular and $| S^{-1}(z) |$ is uniformly bounded. Then $S(z)$ can be continued analytically across the real axis in a neighborhood of σ_0.*

Proof. The main step is to show for almost all σ near σ_0 that $S^{-1}(z)$ is strongly continuous as z tends to σ from below along a vertical segment: We have previously shown for almost all σ that $S(\sigma)$ is unitary and that $S(z)$ approaches $S(\sigma)$ strongly as z tends to σ along vertical segments. At such a σ, given any vector m in N there exists another vector n in N such that

$$S(\sigma)n = m ;$$

furthermore, $S(z)n$ tends strongly to m as z approaches σ. Finally we note that

$$S^{-1}(z)m = S^{-1}(z)S(\sigma)n = S^{-1}(z)\{S(z)n + [S(\sigma) - S(z)]n\}$$

$$= n + S^{-1}(z)[S(\sigma) - S(z)]n.$$

If σ lies near enough to σ_0 and z near enough to σ, then $|\ S^{-1}(z)\ |_N$ is uniformly bounded and since $|\ S(\sigma)n - S(z)n\ |_N$ tends to zero, it follows from the above identity that $S^{-1}(z)m$ tends to $n = S^{-1}(\sigma)m$.

Since $|\ S^{-1}(z)\ |_N$ is uniformly bounded in the lower half-plane near σ_0, we infer that $|\ S^{*-1}(\bar{z})\ |_N$ will be uniformly bounded in the upper half-plane near σ_0. According to Lemma 1.2 $S^*(\bar{z})$ has the same properties as $S(z)$ and hence it follows as above that as z tends to σ from above along a vertical segment that $S^{*-1}(\bar{z})$ tends to $S^{*-1}(\sigma)$ strongly for almost all σ near σ_0. Finally since $S(\sigma)$ is unitary almost everywhere we see that $S^{*-1}(\sigma) = S(\sigma)$ almost everywhere. We have therefore shown: $S^{*-1}(\bar{z})$ is a bounded analytic function in that part of the upper half-plane belonging to a neighborhood of σ_0, whose boundary values on the real axis agree with those of $S(z)$. According to the Schwarz reflection principle of function theory, this guarantees that $S^{*-1}(z)$ is the analytic continuation of $S(z)$ in this neighborhood of σ_0. This completes the proof of Lemma 1.3.

Lemma 1.4. *If $S(z)$ can be continued analytically in a neighborhood of σ_0, then all functions $f(\sigma)$ in R are necessarily analytic in the neighborhood of σ_0.*

Proof. By Lemma 1.1, f belongs to R if and only if $f(\sigma)$ belongs to A_- and $S^*(\sigma)f(\sigma)$ belongs to A_+. Let $g(z)$ denote the function analytic in the upper half-plane which equals $S^*(\sigma)f(\sigma)$ for real z. Then $S^*(\bar{z})f(z)$ is an analytic continuation of $g(z)$ into the lower half-plane near σ_0 by the Schwarz reflection principle. This shows that $g(z)$ is analytic near σ_0 ; but then so is $f(z) = S(z)g(z)$.

2. On Semigroups of Contraction Operators

The fact that the semigroup $\{Z(t)\}$ can be represented as translation followed by truncation acting on a translation invariant subspace is not as special as one might suppose. It turns out that every semigroup of contraction operators can be so represented provided it tends strongly to zero as t becomes infinite. We shall prove this in the present section, which may be skipped for the sake of continuity.

Theorem 2.1. *Let* $\{S(t); t \geq 0\}$ *be a strongly continuous semigroup of contraction operators on a Hilbert space* K *which tends to zero strongly:*

$$\lim_{t \to \infty} S(t)x = 0 ,$$

for every x *in* K. *Then* K *can be represented isometrically as the orthogonal complement of a translation invariant subspace of* $L_2(-\infty, 0; N)$, N *some auxiliary Hilbert space, so that* $S(t)$ *corresponds to right translation by* t *units followed by restriction to* $(-\infty, 0)$.

Remark. The discrete analog of this result is entirely trivial.

Proof. We shall represent the vector x of K by the function of t; $S(-t)x$, $t \leq 0$. This is clearly a translation representation. What remains to construct is a new norm in K, say $|x|_N$, so that the L_2 norm of the function $S(-t)x$ equals the K norm of x, that is

(2.1) $$\int_{-\infty}^{0} |S(-t)x|_N^2 \, dt = |x|_K^2 .$$

We apply this to $S(s)x$ in place of x:

(2.1)' $$|S(s)x|_K^2 = \int_{-\infty}^{0} |S(-t)S(s)x|_N^2 \, dt$$

$$= \int_{-\infty}^{0} |S(-t+s)x|_N^2 \, dt$$

$$= \int_{-\infty}^{-s} |S(-t)x|_N^2 \, dt .$$

Finally differentiating both sides with respect to s at $s = 0$ gives

(2.2) $$|x|_N^2 = -(Bx, x)_K - (x, Bx)_K , \qquad x \text{ in } D_B ,$$

where B is the infinitesimal generator of $\{S(t)\}$.

We now take (2.2) to be the definition of $|x|_N^2$ for all x in the domain of B. Since we have assumed that the operators $S(s)$ are contractions, it follows that the left side of (2.1)' is a nonincreasing function of s; therefore, the negative of its derivative, that is the new norm, is nonnegative.

We can now reverse our steps and verify directly that with the above definition of the new norm (2.1) holds for all x in $D(B)$. For then $S(-t)x$ belongs to $D(B)$ for all values of t, and

$$| S(-t)x |_{\text{new}}{}^2 = -2 \operatorname{Re} (BS(-t)x, S(-t)x)_K = \frac{d}{dt} | S(-t)x |_K{}^2.$$

Integrating with respect to t and using the assumption that $| S(t)x |_K$ tends to zero as t tends to infinity we obtain (2.1).

As the last step we define N as the completion of $D(B)$ in the new norm, more precisely of $D(B)$ modulo the null vectors. Since $D(B)$ is dense in K, we can extend the representation isometrically to all of K.

In this representation the image K_t of K is a right-translation invariant subspace of $L_2(-\infty, 0; N)$. According to a generalization due to Lax [3] (see also Halmos [1]) of a theorem of Beurling [1], such spaces K_t can be represented in the following fashion:

There exists an operator-valued function $S(z)$, analytic and of norm less than or equal to one in the lower half-plane and isometric on the real axis, such that the Fourier transform $\hat{K}$ of K_t is the orthogonal complement of SA_- with respect to A_-, that is

$$\hat{K} = A_- \ominus SA_-.$$

In case $\{S(t)\}$ is the semigroup $\{Z(t)\}$ defined at the beginning of this chapter, S is the previously defined scattering matrix. It is natural therefore to call S the scattering matrix associated with the general semigroup $\{S(t)\}$.

Note. The scattering matrix $S(z)$ is unitary for almost all real z if and only if the translates of K_t are dense in $L_2(-\infty, \infty; N)$.

3. Spectral Theory

Basically both the semigroup $\{Z(t)\}$ and the scattering matrix $S(z)$ are determined by the representation of K in the outgoing translation representation of $\{U(t)\}$. It is not surprising therefore that there is a close relation between the spectral properties of $\{Z(t)\}$ and the behavior of the function $S(z)$.

We recall the following terminology: A closed linear operator V is *regular* if it has a bounded inverse or, equivalently, if it is one-to-one, and onto the whole space. Otherwise it is called singular. The *resolvent set* $\rho(V)$ of V consists of those complex numbers λ for which $\lambda I - V$ is regular; the spectrum $\sigma(V)$ of V consists of those λ for which $\lambda I - V$ is singular. λ belongs to the *point spectrum* of V if $\lambda I - V$ is not one-to-one, that is if its null space contains nonzero vectors.

We shall denote the infinitesimal generator of $\{Z(t)\}$ by B. As the formula (see Appendix 1)

$$(3.1) \qquad (\mu I - B)^{-1} = \int_0^\infty e^{-\mu t} Z(t)\, dt$$

shows, the resolvent set of B contains the right half-plane. As for the left half-plane, we have:

Theorem 3.1. *If* $\operatorname{Re}\mu < 0$, *then* μ *belongs to the point spectrum of B if and only if* $S^*(i\bar\mu)$ *has a nontrivial null space.*

Corollary 3.1. *If* $\operatorname{Re}\mu < 0$, *then* $\bar\mu$ *belongs to the point spectrum of B^* if and only if* $S(i\bar\mu)$ *has a nontrivial null space.*

Proof. Let x be an eigenfunction of B:

$$Bx = \mu x .$$

Then $Z(t)x$ satisfies the differential equation

$$\frac{d}{dt} Z(t)x = B Z(t)x = Z(t)Bx = \mu Z(t)x$$

from which it follows that

$$(3.2) \qquad Z(t)x = e^{\mu t}x .$$

In the outgoing translation representation B acts like differentiation and its domain contains only functions which are continuous for $s \leq 0$. If the eigenvector x has $f(s)$ as its representative, then it follows from (3.2) that f satisfies the functional equation

$$f(s - t) = e^{\mu t}f(s) \qquad (s \leq 0 \leq t) ;$$

the converse assertion is also valid. Making use of the fact that $f(s)$ is continuous for $s \leq 0$ and setting $s = 0$ we obtain that

$$f(s) = \begin{cases} e^{-\mu s}n & (s \leq 0) \\ 0 & (s > 0) \end{cases}$$

for some nonzero n in N. The spectral representation $h(\sigma)$ of x is obtained by taking the Fourier transformation of f. Thus,

$$h(\sigma) = \frac{n}{i\sigma - \mu}.$$

It will be recalled that K corresponded in the outgoing spectral representation to the function space R. According to Lemma 1.1 a function h in A_- belongs to R if and only if $\mathcal{S}^* h$ belongs to A_+; by the Paley–Wiener theorem this will be the case if and only if $\mathcal{S}^* h$ can be continued analytically into the upper half-plane and the integrals of its absolute value squared along the lines Im $z = c > 0$ are uniformly bounded. Now $\mathcal{S}^* h$ has a unique meromorphic continuation into the upper half-plane, namely

(3.3)
$$\frac{\mathcal{S}^*(\bar{z})n}{iz - \mu}.$$

This will be analytic if and only if the numerator vanishes at the zero $z = -i\mu$ of the denominator, that is if and only if

(3.4)
$$\mathcal{S}^*(i\bar{\mu})n = 0.$$

If so, the function (3.3) is clearly uniformly square integrable along the lines Im $z = c > 0$. This proves the theorem, and a little more:

Corollary 3.2. $\dim[null\ space\ of\ (\mu I - B)] = \dim[null\ space\ of\ \mathcal{S}^*(i\bar{\mu})]$.

Remark. If Re $\mu = 0$ then μ can not belong to the point spectrum of B since $z(t)x = e^{\mu t}x$ does not in this case converge to zero as $t \to \infty$.

We shall need the following more general result:

Theorem 3.2. (i) *If* Re $\mu < 0$, *then* μ *belongs to the resolvent set of* B *if and only if* $\mathcal{S}(i\bar{\mu})$ *is regular.* (ii) *A purely imaginary* μ_0 *belongs to the resolvent set of* B *if and only if* $\mathcal{S}(z)$ *can be continued analytically across the real axis at* $\sigma_0 = i\bar{\mu}_0$.

Proof. We shall first prove the if parts of these two assertions. In the case of (i) we assume for some μ with Re $\mu < 0$ that $\mathcal{S}^*(i\bar{\mu})$ is regular and we wish to show that $\mu I - B$ is one-to-one and onto K. Theorem 3.1 implies that this operator is one-to-one and to prove that its range is K we proceed as follows:

As already noted B acts like $-d/ds$ in the outgoing translation representation and its domain consists of functions with support in $(-\infty, 0]$ whose derivative over $(-\infty, 0)$ is square integrable. Taking the Fourier transform we obtain the following description of B and its domain in the spectral representation:

Lemma 3.1. *A function $h(\sigma)$ in R belongs to the domain of B if and only if there exists a vector n in N such that*

$$(3.5) \qquad\qquad i\sigma h(\sigma) - n$$

belongs to R; Bh is given by (3.5).

Note. The sufficiency argument for Lemma 3.1 is a straightforward application of distribution theory. Thus, if the inverse Fourier transforms of h and Bh are $f(s)$ and $g(s)$, respectively, the Plancherel theorem gives

$$\int g\phi \, ds = \int f\phi' \, ds - n\phi(0)$$

for all test functions $\phi(s)$ and the assertion follows directly from this.

Thus, in order to prove that $\mu I - B$ maps its domain onto K it suffices to take an arbitrary k in R and to show that there exists an n in N and an h in R such that

$$(\mu - i\sigma)h(\sigma) + n = k(\sigma) \ .$$

Clearly,

$$(3.6) \qquad\qquad h(\sigma) = \frac{k(\sigma) - n}{\mu - i\sigma} \ ;$$

what remains to be shown is that for suitable n, h belongs to R.

Now, according to Lemma 1.1, h belongs to R if and only if h belongs to A_- and S^*h belongs to A_+. By the Paley–Wiener theorem h lies in A_- if and only if it can be extended to be analytic in the lower half-plane and the integral of its absolute value squared along the lines $\operatorname{Im} z = c < 0$ are uniformly bounded. The function k already lies in R so it can be so extended, and since the denominator in (3.6) does not vanish in the lower half-plane, it follows from (3.6) that

$$\frac{k(z) - n}{\mu - iz}$$

is the required extension of h; so h belongs to A_- for any choice of n.

We next consider S^*h:

$$S^*(\sigma)h(\sigma) = \frac{S^*(\sigma)k(\sigma) - S^*(\sigma)n}{\mu - i\sigma}.$$

Since k belongs to R, by Lemma 1.1, S^*k can be extended to a uniformly square integrable analytic function $g(z)$ in the upper half-plane. Thus, S^*h has the *meromorphic* continuation

$$(3.7) \qquad \frac{g(z) - S^*(\bar{z})n}{\mu - iz}$$

into the upper half-plane. This will be analytic if and only if the numerator in (3.7) vanishes at the point $z = -i\mu$ where the denominator vanishes, that is if and only if

$$(3.8) \qquad g(-i\mu) = S^*(i\bar{\mu})n.$$

Since we have assumed that $S(i\bar{\mu})$ and hence $S^*(i\bar{\mu})$ is regular, Eq. (3.8) determines n uniquely. With such a choice for n the function (3.7) is analytic and, as may be easily checked, the integrals of its absolute value squared along the lines $\operatorname{Im} z = c > 0$ are uniformly bounded. So by the Paley–Wiener theorem S^*h belongs to A_+ and thus by Lemma 1.1 we see that h belongs to R. This completes the if part of Theorem 3.2 (i).

Our construction of $(\mu I - B)^{-1}$ is quite explicit and can be used to obtain an estimate for the norm of the resolvent which we will require later on. The relation (3.6) and the triangle inequality give

$$|h| \le \left| \frac{k}{\mu - i\sigma} \right| + \left| \frac{n}{\mu - i\sigma} \right|.$$

The first term on the right can be estimated by replacing the denominator by its smallest value; the second term can be evaluated explicitly. Using the abbreviation $\tau = -\operatorname{Re} \mu$, we obtain

$$(3.9) \qquad |h| \le \tau^{-1} |k| + (2\tau)^{-1/2} |n|_N.$$

The value of n is given by (3.8):

$$n = \mathbb{S}^{*-1}(i\bar{\mu})g(-i\mu) \; ;$$

and hence,

(3.10) $$| n |_N \le | \mathbb{S}^{-1}(i\bar{\mu}) |_N | g(-i\mu) |_N .$$

Now $g(-i\mu)$ can be expressed by the Cauchy integral and the resulting integral can be estimated by the Schwarz inequality. This gives

(3.11) $$| g(-i\mu) |_N = \frac{1}{2\pi} \left| \int_{-\infty}^{\infty} \frac{g(\sigma)}{\sigma + i\mu} \, d\sigma \right|_N \le (2\tau)^{-1/2} | g | .$$

Moreover, since $\mathbb{S}^*(\sigma)$ is unitary on the real axis,

$$| g | = | \mathbb{S}^* k | = | k | ,$$

and substituting this into (3.11), (3.10), and (3.9) successively we obtain

$$| h | \le \tau^{-1} | k | + (2\tau)^{-1} | \mathbb{S}^{-1}(i\bar{\mu}) |_N | k |$$

$$\le (3/2\tau) | \mathbb{S}^{-1}(i\bar{\mu}) |_N | k | .$$

Finally, since $(\mu I - B)h = k$, we obtain the inequality

(3.12) $$| (\mu I - B)^{-1} | \le \frac{3 | \mathbb{S}^{-1}(i\bar{\mu}) |_N}{2 | \operatorname{Re} \mu |} .$$

The same analysis which applied to part (i) also applies to part (ii) where the assumption is that $\mathbb{S}(z)$ is analytic at $\sigma_0 = i\bar{\mu}_0$. In this case each function $k(\sigma)$ in R is analytic at σ_0 by Lemma 1.4. Using this fact it is easy to verify that if n is chosen as $k(\sigma_0)$, then h, given by (3.6), belongs to A_- and $\mathbb{S}^* h$ belongs to A_+.

Next we prove the only if part of Theorem 3.2, starting with part (i). Thus for $\operatorname{Re} \mu < 0$ we now assume that μ belongs to $\rho(B)$ and wish to show that $\mathbb{S}(i\bar{\mu})$ is regular. We shall argue by contradiction and assume that $\mathbb{S}(i\bar{\mu})$ is not regular. Then neither is $\mathbb{S}^*(i\bar{\mu})$ and one of two possibilities must hold:

   (a) $\mathbb{S}^*(i\bar{\mu})$ is not one-to-one;
   (b)   The range of $\mathbb{S}^*(i\bar{\mu})$ is not all of K.

In the case of (a), μ would belong to the point spectrum of B by Theorem 3.1. This is contrary to μ belonging to $\rho(B)$. Case (b) can be divided into the usual two subclasses:

(b)′   The range of $S^*(i\bar{\mu})$ is not dense in K;
(b)″ $S^{*-1}(i\bar{\mu})$ is unbounded.

In case (b)′ the null space of $S(i\bar{\mu})$ would be nontrivial and so, according to Corollary 3.1, μ would belong to the point spectrum of B, again contradicting the assumption that μ lies in $\rho(B)$.

It remains to show that (b)″ is impossible and for this we need the following inequality:

Lemma 3.2.   *Given any μ with* $\mathrm{Re}\,\mu < 0$ *and any unit vector n in N, there exists a nonzero vector a in K such that for all positive t*

$$(3.13) \qquad |\,[Z(t) - e^{\mu t}I]a\,| \leq 4\,|\,S^*(i\bar{\mu})n\,|_N\,|\,a\,|\,.$$

Before presenting a proof for Lemma 3.2 we shall show how to use (3.13) to establish case (b)″. For this purpose we begin with the expression (3.1) for the resolvent of B and a positive κ:

$$(\kappa I - B)^{-1} = \int_0^\infty e^{-\kappa t}Z(t)\,dt.$$

Subtraction $(\kappa - \mu)^{-1}$ from both sides we obtain the identity

$$(3.14) \qquad (\kappa I - B)^{-1} - (\kappa - \mu)^{-1} = \int_0^\infty e^{-\kappa t}[Z(t) - e^{\mu t}]\,dt\,.$$

Let $Q = Q(\mu)$ denote the operator on the left in (3.14). Then Q can be rewritten as

$$Q(\mu) = -(\mu I - B)(\kappa I - B)^{-1}(\kappa - \mu)^{-1}\,.$$

It is clear from this identity that $Q(\mu)$ is regular if and only if $(\mu I - B)$ is one-to-one and onto.

Now, according to Lemma 3.2, given any unit vector n in N there exists a nonzero element a in K such that (3.13) is satisfied. We use (3.14) to express Qa and then estimate the norm of the resulting expression with the aid of (3.13); abbreviating $S^*(i\bar{\mu})$ by S^* we get

$$| Qa | = \left| \int_0^\infty e^{-\kappa t}[Z(t) - e^{\mu t}I]a\, dt \right| \leq \int_0^\infty e^{-\kappa t}4\, | S^*n |_N | a |\, dt$$

$$\leq \frac{4}{\kappa} | S^*n |_N | a |.$$

Since n is a unit vector this last inequality can be rewritten as

(3.15)
$$\frac{| n |_N}{| S^*n |_N} = \frac{4}{\kappa} \frac{| a |}{| Qa |}.$$

We have assumed that $(\mu I - B)$ is regular and hence, as we have noted above, so is Q. This implies that the right side of (3.15) is bounded and therefore that the inverse of S^* is bounded; that is, the case (b)$''$ cannot occur. This completes the proof of the regularity of $S^*(i\bar{\mu})$.

The inequality (3.15) can be expressed thus:

(3.15)$'$
$$| S^{*-1}(i\bar{\mu}) |_N \leq \frac{4}{\kappa} | Q^{-1}(\mu) |$$

at all points N of $\rho(B)$ in Re $\mu < 0$.

Finally we address ourselves to the only if part of Theorem 3.2 (ii): We have to show that if μ_0 is pure imaginary and belongs to $\rho(B)$, then $S(z)$ is analytic at $\sigma_0 = i\bar{\mu}_0$. Since the resolvent set is open, all points μ sufficiently close to μ_0 also belong to $\rho(B)$ and hence, as above, $Q(\mu)$ will be regular for all such μ and

$$Q^{-1}(\mu) = -(\kappa I - B)(\mu I - B)^{-1}(\kappa - \mu)$$

(3.16)
$$= (\mu - \kappa)I - (\kappa - \mu)^2(\mu I - B)^{-1}$$

is uniformly bounded in norm in a neighborhood of μ_0. According to the inequality (3.15)$'$ this implies that $S^{-1}(z)$ is uniformly bounded for all z in the lower half-plane which are close enough to $\sigma_0 = i\bar{\mu}_0$. Furthermore, Lemma 1.3 guarantees when this is the case that $S(z)$ is analytic at σ_0. Our proof of Theorem 3.2 is now complete modulo the proof of Lemma 3.2.

In order to establish Lemma 3.2 we have for a given n in N to construct an element a in K such that the inequality (3.13) holds. We recall from the proof of Theorem 3.1 that if n were such that $S^*(i\bar\mu)n = 0$ then we could choose the element a to be the corresponding eigenvector for B and hence for $Z(t)$; in this case the outgoing translation representation of a would be $f(s) = \exp(-\mu s)n$ for $s \leq 0$ and $= 0$ for $s > 0$. We shall make the same choice for general n but since in this case the above exponential function does not represent an element in K, we take for a the nearest thing to it in K, namely its orthogonal projection into K.

We shall operate in the outgoing spectral representation where the exponential function is turned by Fourier transformation into

$$e = e(n, \mu) = M \frac{n}{i\sigma - \mu}, \qquad M = (2 \mid \operatorname{Re} \mu \mid)^{1/2};$$

the normalizing factor M is chosen so that

$$\mid e \mid = \mid n \mid_N.$$

Next we decompose e as

(3.17) $$e = a + b,$$

a in R, b orthogonal to R but in A_- ; that is, b lies in the image SA_- of D_- .

It turns out that a and b can be determined explicitly; to do this we multiply (3.17) by $S^*(\sigma)$:

(3.18) $$S^*e = S^*a + S^*b.$$

As we have already remarked, b is of the form Sc for some c in A_- . Since S is unitary, $S^*S = I$ and so $S^*b = c$. On the other hand Lemma 1.1 asserts that S^*a belongs to A_+ if a is in R. Thus, (3.18) is the orthogonal decomposition of $S^*e = M(i\sigma - \mu)^{-1}S^*(\sigma)n$ into its A_+ and A_- components. This decomposition can be seen by inspection to be:

(3.19) $$M \frac{S^*(\sigma)n}{i\sigma - \mu} = M \frac{S^*(\sigma)n - S^*(i\bar\mu)n}{i\sigma - \mu} + M \frac{S^*(i\bar\mu)n}{i\sigma - \mu}.$$

The first term on the right is the boundary value of $M(iz - \mu)^{-1}[S^*(\bar z)n - S^*(i\bar\mu)n]$ and belongs to A_+ by the Paley–Wiener theorem; it is obvious that the second term lies in A_- .

The norm of S^*b can be computed from (3.19) to be $\mid S^*(i\bar\mu)n \mid_N$ and since S^* is unitary, this is equal to the norm of b, that is,

$$(3.20) \qquad |b| = |S^*(i\bar{\mu})n|_N .$$

By the Pythagorean relation and (3.17) we have

$$(3.21) \qquad |a|^2 = |n|_N^2 - |S^*(i\bar{\mu})n|_N^2 .$$

It is clear from the outgoing translation representation of $\{U(t)\}$ that the exponential function is an eigenfunction of $P_+U(t)$, that is, $P_+U(t)e = \exp(\mu t)e$. Applying $[P_+U(t) - \exp(\mu t)I]$ to (3.17) we get

$$(3.22) \qquad 0 = [Z(t) - e^{\mu t}I]a + [P_+U(t) - e^{\mu t}I]b .$$

Obviously $|P_+U(t) - \exp(\mu t)I| \leq |P_+U(t)| + |\exp(\mu t)| \leq 2$; and so we conclude from (3.22), using (3.20), that

$$(3.23) \quad |[Z(t) - e^{\mu t}I]a| = |[P_+U(t) - e^{\mu t}I]b| \leq 2|b| = 2|S^*(i\bar{\mu})n|_N .$$

We have trivially that

$$(3.24) \qquad |[Z(t) - e^{\mu t}I]a| \leq 2|a| .$$

Take n to be a unit vector. If $|S^*(i\bar{\mu})n|_N \geq 1/2$, then (3.24) implies the desired inequality (3.13); if $|S^*(i\bar{\mu})n|_N < 1/2$, then it follows from (3.21) that $|a| > 1/2$ and so we have from (3.23) that

$$|[Z(t) - e^{\mu t}I]a| \leq 2|S^*(i\bar{\mu})n|_N \leq 4|S^*(i\bar{\mu})n|_N|a| ,$$

which is again the inequality (3.13). This completes the proof of Lemma 3.2 and thereby that of Theorem 3.2.

4. A Spectral Mapping Theorem

The next result that we need is a spectral mapping theorem known from the theory of semigroups of operators (see Phillips [1]). Although it can be derived in full generality with the aid of the Gelfand representation theorem for commutative Banach algebras, we shall give a proof for the theorem in this special setting for the sake of completeness and also because it is instructive and amusing to derive it with the aid of the scattering matrix.

We begin with a few words about the functional calculus for the infinitesimal generator B of the semigroup $\{Z(t)\}$. Let $m(dt)$ be any complex measure on the Borel subsets of the nonnegative reals with total measure finite. Its Laplace transform

$$(4.1) \qquad g(\mu) = \int_0^\infty e^{\mu t} m(dt), \qquad \mathrm{Re}\, \mu \le 0$$

is analytic in the left half-plane and continuous up to the imaginary axis.

Remark. The Laplace transforms of measures form an algebra under multiplication, the product of two transforms being the transform of the convolution of the corresponding measures.

Definition 4.1. If $g(\mu)$ is a function of the form (4.1), we define $g(B)$ as

$$(4.2) \qquad g(B) = \int_0^\infty Z(t) m(dt) .$$

This correspondence defines a homomorphism of Laplace transforms into the algebra of linear bounded operators:

Lemma 4.1. *If $g_i(B)$ corresponds to the measure $m_i(dt)$ as in (4.2) for $i = 1, 2, 3$ and if $m_3 = m_1 * m_2$, then*

$$g_3(B) = g_1(B) g_2(B) .$$

Proof. This result follows directly from the identity

$$g_3(B) = \int_0^\infty Z(r) [m_1 * m_2](dr)$$

$$= \int_0^\infty \int_0^\infty Z(t+s) m_1(dt) m_2(ds)$$

$$= g_1(B) g_2(B) .$$

Next we prove the spectral mapping theorem:

Theorem 4.1. *If μ_0 belongs to the spectrum of B, then $g(\mu_0)$ belongs to the spectrum of $g(B)$.*

Proof. As we have seen, $\mathrm{Re}\, [\sigma(B)] \le 0$. Suppose first that $\mathrm{Re}\, \mu_0 < 0$; then according to Theorem 3.2 the operator $S^*(i\bar{\mu}_0)$ is singular. As in the proof of Theorem 3.2 there are three possibilities:

(a) $S^*(i\bar{\mu}_0)$ is not one-to-one:

(b)′   The range of $S^*(i\bar{\mu}_0)$ is not dense in N;

(b)″ $S^{*-1}(i\bar{\mu}_0)$ is unbounded.

If (a) [or (b)′] is valid then it follows from Theorem 3.1 that B [or B^*] has an eigenfunction with eigenvalue $\mu_0[\bar{\mu}_0]$. This eigenfunction corresponds to an exponential (with support $(-\infty, 0]$ [or $[0, \infty)$]) in the outgoing [incoming] translation representation which shows that it is also an eigenfunction of $g(B)$ [or $g(B)^*$] with eigenvalue $g(\mu_0)$ [or $\overline{g(\mu_0)}$].† This proves the assertion in case (a) or (b)′ holds.

Next we show that (b)″ is incompatible with the regularity of $[g(B) - g(\mu_0)I]$. By definition

$$g(B) - g(\mu_0)I = \int_0^\infty [Z(t) - \exp{(\mu_0 t)}I]m(dt) .$$

According to Lemma 3.2, given any unit vector n in N there exists a nonzero element a in K such that the inequality (3.13) is satisfied. The expression $|[g(B) - g(\mu_0)I]a|$ can be estimated with the aid of this inequality as

$$|[g(B) - g(\mu_0)I]a| \le 4M |S^*(i\bar{\mu}_0)n|_N |a|, \qquad M = \int_0^\infty |m(dt)|.$$

Since n is a unit vector this can be rewritten as

(4.3) $$\frac{|n|_N}{|S^*(i\bar{\mu}_0)n|_N} \le 4M \frac{|a|}{|[g(B) - g(\mu_0)I]a|}.$$

If $[g(B) - g(\mu_0)I]$ is regular then the right side is bounded for all a and the left side will be bounded for all n in N, contradicting (b)″. We note that the relation (4.3) can be expressed as:

(4.4) $$|S^{-1}(i\bar{\mu}_0)|_N \le 4M |[g(B) - g(\mu_0)I]^{-1}|.$$

Finally, if Re $\mu_0 = 0$ and $[g(B) - g(\mu_0)I]$ is regular, then since the resolvent set for $g(B)$ is open and the resolvent itself continuous, and since $g(\mu)$ is continuous up to the imaginary axis, there will be a neighbor-

† Note that $g(B)^* = h(B^*)$ where

$$h(N) = \int_0^\infty e^{\mu t}\bar{m}(dt) = \overline{g(\bar{\mu})}.$$

hood about μ_0 such that $| [g(B) - g(\mu)]^{-1} |$ is bounded for all μ with
Re $\mu \leq 0$ in this neighborhood. According to (4.4) $| S^{-1}(i\bar{\mu}) |$ will be
bounded in this same set and consequently by Lemma 1.3 $S(z)$ will be
analytic in a neighborhood of $i\bar{\mu}_0$. In this case Theorem 3.2 (ii) applies
and we see that $[\mu_0 I - B]$ is necessarily regular. This concludes the proof
of Theorem 4.1.

From (4.4) we deduce:

Corollary 4.1. *If λ belongs to $\rho[g(B)]$, then $S(z)$ is regular and its
inverse uniformly bounded at all points $i\bar{\mu}$ for which $g(\mu) = \lambda$ and Re $\mu \leq 0$.*

It might be interesting to investigate the class of functions g for which
the above necessary condition for λ to belong to $\rho[g(B)]$ is also sufficient.
It is not difficult to show that it is true for $g(B) = \exp(Bt) = Z(t)$; this
was shown by Moeller [1] for the scalar case and the proof extends to the
vector-valued case. Since we have no use for results of this kind in the
present monograph, we shall not pursue the subject further.

On the other hand the following result plays an important part in our
theory.

Theorem 4.2. *If for some function g analytic on the spectrum of B and
of the form (4.1) the operator $g(B)$ is compact then B has a pure point spectrum
except possibly for those values of μ_0 for which $g(\mu_0) = 0$. The spectrum of B
can have only such points as points of accumulation.*

Proof. Suppose that μ is in $\sigma(B)$ and set $\lambda = g(\mu)$. Then by the spectral
mapping theorem λ belongs to $\sigma[g(B)]$. Since $g(B)$ is a compact operator
its spectrum consists of a denumerable set of points with zero as the only
possible point of accumulation. If $g(\mu) = $ const then $g(B) = $ const. I can
be compact only when K is of finite dimension in which case the theorem
is trivially true. Otherwise g is nonconstant and analytic on the spectrum
of B and hence the equation

$$(4.5) \qquad\qquad g(\mu) = \lambda$$

has at most a denumerable number of solutions in $\sigma(B)$; this shows that
the spectrum of B is denumerable. Further, if the sequence $\{\mu_n\}$ belongs
to $\sigma(B)$ and converges to μ_0 then since g is analytic an infinite subset of the
$g(\mu_n)$ are distinct so that $g(\mu_0)$ is a point of accumulation of the spectrum
of $g(B)$ and hence $g(\mu_0) = 0$. This shows that the spectrum of B is a de-

numerable point set whose only points of accumulation are at those values of μ_0, for which $g(\mu_0) = 0$.

It remains to show that the spectrum of B is a pure point spectrum except possibly for those values μ_0 for which $g(\mu_0) = 0$. Now for $\lambda \neq 0$ in the spectrum of $g(B)$, the operator $[\lambda I - g(B)]$ has a finite dimensional null space N_λ. The operators $Z(t)$ commute with $g(B)$ and therefore N_λ is an invariant subspace for $Z(t)$ (Schur's lemma). According to the elementary spectral theory for semigroups of operators over a finite dimensional space, N_λ belongs to the domain of B and is an invariant subspace of B. Consequently N_λ contains an eigenvector x of B:

$$Bx = \nu x .$$

Clearly, $Z(t)x = \exp(\nu t)x$ and so $g(B)x = g(\nu)x$. Since N_λ consists of all eigenvectors of $g(B)$ for the eigenvalue λ, it follows that

$$(4.6) \qquad\qquad g(\nu) = \lambda .$$

Let κ be any point in the right half-plane and define the function g_κ by

$$g_\kappa(\mu) = \frac{1}{\kappa - \mu} g(\mu) .$$

The function g_κ is the product of two functions each analytic on the spectrum of B and of the form (4.1) so that g_κ is itself of that form and $g_\kappa(B) = (\kappa I - B)^{-1}g(B)$. The operator $g_\kappa(B)$ is also compact since it is the product of two bounded operators of which one, namely $g(B)$, is compact. Therefore the results derived above apply to the function g_κ as well. Thus, for μ in $\sigma(B)$, $g_\kappa(\mu)$ belongs to $\sigma[g_\kappa(B)]$ and there exists an eigenvalue ν_κ in the point spectrum of B such that

$$g_\kappa(\nu_\kappa) = g_\kappa(\mu) .$$

Using the definition of g_κ and the abbreviation λ for $g(\mu)$ we can readily derive from the above that

$$(4.7) \qquad\qquad g(\nu_\kappa) = \frac{\kappa - \nu_\kappa}{\kappa - \mu} \lambda .$$

Since ν_κ is an eigenvalue of B and since the spectrum of B, as shown above consists of a denumerable set of points, there must be a value of ν

which equals ν_κ for a nondenumerable set of κ. But then it follows from (4.7) that

$$\frac{\kappa - \nu}{\kappa - \mu}\lambda$$

is independent of κ on this nondenumerable set and since $\lambda \neq 0$ we conclude that $\nu = \mu$. Thus we have shown that μ belongs to the point spectrum of B.

Actually we can prove a bit more under the assumptions of Theorem 4.2 for points μ in $\sigma(B)$ such that $g(\mu) \neq 0$:

Corollary 4.2.   Re $\mu < 0$.

Corollary 4.3.   *The null space of $\mu I - B$ is finite dimensional.*

Corollary 4.4.   *The resolvent of B has a pole at μ.*

Corollary 4.5.   *If $g(\mu)$ has no zero on the spectrum of B, then the resolvent of B is meromorphic in the whole plane.*

Since μ lies in the point spectrum of B by the previous theorem, there is an eigenvector x such that $Z(t)x = \exp(\mu t)x$. In view of the fact that $Z(t)$ tends strongly to zero it is impossible for Re μ to be zero; this proves Corollary 4.2. Corollary 4.3 follows from the fact that the null space of $\mu I - B$ is a subspace of the null space of $[g(\mu)I - g(B)]$, and the latter is finite dimensional.

Proof of Corollary 4.4. According to Theorem 4.2, μ is an isolated point in $\sigma(B)$ and hence an isolated singularity of $(\mu I - B)^{-1}$. It remains to show that it is a pole. Since $g(B)$ is compact, its resolvent has a pole at $\lambda \equiv g(\mu) \neq 0$; let m be the order of the pole. Then, for ν near enough but different from λ,

$$(4.8) \qquad |[\gamma I - g(B)]^{-1}| \leq \text{const} \, |\gamma - \lambda|^{-m}.$$

Since g is analytic near μ, every point γ near λ can be expressed as

$$\gamma = g(\nu)$$

with ν near μ, possibly in more than one way. If the derivative of g has a zero of order $p - 1$ then

$$|\nu - \mu|^p \leq \text{const} \, |\gamma - \lambda|.$$

Inserting this into (4 8) gives

$$| [\gamma I - g(B)]^{-1} | \leq \text{const} \, | \nu - \mu |^{-mp} ,$$

and combining this with the inequality (4.4) (recall that $\text{Re} \, \mu < 0$) gives

$$| S^{-1}(i\bar{\nu}) | \leq \text{const} \, | \nu - \mu |^{-mp} .$$

Substituting this estimate for S^{-1} into the inequality (3.12) yields finally

$$| (\nu I - B)^{-1} | \leq \text{const} \, | \nu - \mu |^{-mp}$$

as asserted in Corollary 4.4.

Proof of Corollary 4.5. If $g(\mu)$ is nowhere zero on the spectrum of B then the latter has at most infinity as its only point of accumulation, and by the foregoing corollaries the imaginary axis belongs to the resolvent set and each point of the spectrum is a pole for the resolvent. Consequently $(\nu I - B)^{-1}$ is meromorphic in the whole plane as asserted.

Remark 4.1. In the course of the proof of Theorem 4.2 we have shown that if λ belongs to the spectrum of $g(B)$, $\lambda \neq 0$, then there is a ν in the spectrum of B such that $\lambda = g(\nu)$. If $g(B)$ is not compact, this is not necessarily true, see Corollary 4.1.

5. Applications of the Spectral Theory

The results of the preceding two sections relate the behavior of the scattering matrix $S(z)$ to the spectral properties of the semigroup of operators $\{Z(t)\}$. This opens up two possibilities: One is to study directly the behavior of the scattering matrix in the lower half-plane and thereby obtain information about the semigroup; the other is to study $\{Z(t)\}$ directly and thereby obtain information about the scattering matrix. In this monograph we shall follow the second course. The direct study of $\{Z(t)\}$ for the wave equation will be carried out in Chapter V; in preparation for that we state and prove in this section some results, all of them simple consequences of Sections 3 and 4.

Theorem 5.1. *If for some positive values of T and κ, the operator $Z(T)(\kappa I - B)^{-1}$ is compact, then B has a pure point spectrum and the scattering matrix $S(z)$ is holomorphic on the real axis and meromorphic in the whole plane, having a pole at each point z for which iz belongs to the spectrum of B.*

Theorem 5.2. *If for some value T, $|Z(T)| < \alpha < 1$, then $S(z)$ is holomorphic and bounded in the strip $\operatorname{Im} z < -(\log \alpha)/T$. Conversely, if $S(z)$ is holomorphic and bounded in the strip $\operatorname{Im} z < \gamma$, then*

$$(5.1) \qquad \lim_{t \to \infty} t^{-1} \log |Z(t)| \le -\gamma.$$

Theorem 5.3. *If for some value of T, $Z(T)$ is compact, then the scattering matrix is meromorphic in the whole plane and each horizontal strip contains only a finite number of its poles.*

Proof of Theorem 5.1. If we apply Theorem 4.2 and Corollaries 4.2 and 4.5 to the function $g(\mu) = \exp(\mu T)(\kappa - \mu)^{-1}$, we see that B has a pure point spectrum and that the resolvent of B is holomorphic on the imaginary axis and meromorphic in the whole plane. Hence, by Theorem 3.2 the scattering matrix $S(z)$ can be continued analytically across the real axis, its analytic continuation being equal to $S^{*-1}(\bar{z})$ which is analytic at all points z at which $S(\bar{z})$ is regular. Finally we see from (3.12) that $S^{-1}(\bar{z})$ will be singular at each spectral point iz of B and from (3.15)′ and (3.16) that the singularities are no worse than those of $(izI - B)^{-1}$.

Proof of Theorem 5.2. Since $|Z(T)| = \alpha_0 < \alpha < 1$ we have for every positive integer n

$$(5.2) \qquad |Z(nT)| = |Z^n(T)| \le |Z(T)|^n \le \alpha_0{}^n = e^{-n\beta T},$$

β being defined as $-(\log \alpha_0)/T$. Let t be any positive number and decompose it mod T, that is, write it as $t = nT + r$, n a nonnegative integer and $0 \le r < T$. From (5.2) and the fact that $|Z(r)| \le 1$ we get

$$|Z(t)| = |Z(nT)Z(r)| \le |Z(nT)| \, |Z(r)| \le ce^{-\beta t},$$

where $c = \exp(\beta T)$; in other words $Z(t)$ decays exponentially. Using this estimate for $|Z(t)|$ in the Laplace transform expression (3.1) for the resolvent of B, we obtain the result that $(\mu I - B)^{-1}$ is holomorphic for $\operatorname{Re} \mu > -\beta$ and bounded in norm in any smaller strip. According to Theorem 3.2 this implies that the scattering matrix is regular for $-\beta < \operatorname{Im} z \le 0$ and according to the inequality (3.15)′ its reciprocal is bounded in any smaller strip. Again, by Theorem 3.2, $S(z)$ is holomorphic on the real axis and hence can be continued by reflection to be holomorphic in the strip $0 \le \operatorname{Im} z < \beta$ and bounded in any smaller strip.

To prove the converse assertion let u, v be elements of R. Then v is in A_- and by Lemma 1.1 $u = Sw$ where w is in A_+. Making use of the outgoing spectral representation for $\{U(t)\}$ we can write

$$(5.3) \qquad (Z(t)u, v) = (U(t)u, v) = \int_{-\infty}^{\infty} e^{i\sigma t}(Sw, v)_N \, d\sigma.$$

By hypothesis the integrand can be extended analytically into the strip $\operatorname{Im} z < \gamma$ as $e^{izt}(S(z)w(z), v(\bar{z}))_N$ and in this strip $|\,S(z)\,|_N \leq M_\gamma$. Choose $\gamma_0 < \gamma$. It is easy to see that we can transform the path of integration to the line $\operatorname{Im} z = \gamma_0$ and applying the above estimate for $|\,S(z)\,|_N$ and the Schwarz inequality to the so-transformed integral, we obtain

$$(5.4) \quad |\,(Z(t)u, v)\,| \leq \exp(-\gamma_0 t) M_\gamma \,|\,w\,|\,|\,v\,| = \exp(-\gamma_0 t) M_\gamma \,|\,u\,|\,|\,v\,|;$$

here we have made use of the fact that the integral of the square of $w(z)$ along the line $\operatorname{Im} z = \gamma_0$ is less than or equal to $|\,w\,|^2$ and the analogous result for v. Since (5.4) holds for each $\gamma_0 < \gamma$, it follows that

$$\limsup t^{-1} \log |\,Z(t)\,| \leq -\gamma.$$

Finally, because of the fact that $\log |\,Z(t)\,|$ is a subadditive function, the limit itself exists.

Proof of Theorem 5.3. The first part of the theorem which states that $S(z)$ is meromorphic in the whole plane follows from Theorem 5.1 since $Z(T)(\kappa I - B)^{-1}$ is compact if $Z(T)$ is compact. To prove the second part we argue indirectly: suppose that there were infinitely many poles $\{\pi_j\}$ in some strip: $\operatorname{Im} \pi_j \leq \gamma$. It follows again from Theorem 5.1 that $i\pi_j$ belongs to the spectrum of B and so by the spectral mapping theorem $\exp(i\pi_j t)$ belongs to the spectrum of $Z(t)$. Choose t so that t is greater than T and so that the numbers $\{\exp(i\pi_j t)\}$ are all distinct (this excludes a denumerable set of values of t). For such a t, $Z(t) = Z(t - T)Z(T)$ is compact and therefore has only a finite number of points in its spectrum with absolute value greater than $\exp(-\gamma t)$. But this contradicts the fact that all of the numbers $\{\exp(i\pi_j t)\}$ belong to the spectrum of $Z(t)$. This completes the proof of the second part of Theorem 5.3.

In the course of proving Theorem 5.2 we have shown that the assumption that $|\,Z(T)\,|$ is less than one for some value of T implies that $|\,Z(t)\,|$ decays exponentially. Next we show that the hypothesis of Theorem 5.3 implies even more.

Theorem 5.4. *If for some value of T, $Z(T)$ is compact then the eigenfunction expansion for $Z(t)$ is asymptotically valid for large t in the following sense:*

Arrange the eigenvalues μ_j of B in decreasing order of their real parts and denote by P_j the projection onto the jth eigenspace (for simplicity we forget about generalized eigenspaces). Then

$$(5.5) \qquad Z(t) \sim \sum \exp(\mu_j t) P_j$$

holds in the sense that for each $\epsilon > 0$

$$(5.6) \qquad |Z(t) - \sum_{j=1}^{n} \exp(\mu_j t) P_j| \leq \text{const} \, | \exp((\mu_{n+1} + \epsilon)t) |,$$

the value of the constant depending on n and ϵ.

We sketch the proof of this well-known result: Let M_n denote the null space of the projection operator $\sum_1^n P_j$. Clearly M_n is an invariant subspace for $\{Z(t)\}$ and the spectral radius of $Z(T)$ over M_n is less than or equal to $|\exp(\mu_{n+1}T)|$. For otherwise there would be an eigenvalue λ with $|\lambda| > |\exp(\mu_{n+1}T)|$ of $Z(T)$ over M_n. By Remark 4.1 there will be an eigenvalue ν and an eigenfunction in M_n for B with $\lambda = \exp(\nu T)$; but this is contrary to our ordering of the spectrum of B. According to the Gelfand formula for the spectral radius we therefore have

$$\lim_{k \to \infty} |Z(T)^k|_n^{1/k} \leq |\exp(\mu_{n+1}T)|,$$

where $|\cdot|_n$ denotes the norm over M_n. Since $Z(T)^k = Z(kT)$ and $|Z(r)| \leq 1$, this shows that for a given ϵ and t sufficiently large

$$(5.7) \qquad |Z(t)|_n \leq \text{const} \, | \exp[(\mu_{n+1} + \epsilon)t]|.$$

For any element u in K, let $u_n = \sum_1^n P_j u$, that is, u_n is the projection of u on the first n eigenspaces. The remainder $r_n = u - u_n$ belongs to M_n. Then u_n depends boundedly on u and therefore so does r_n:

$$(5.8) \qquad |r_n| \leq \text{const} \, |u|.$$

Setting

$$R_n(t) = Z(t) - \sum_1^n \exp(\mu_j t) P_j,$$

it is clear that $R_n(t)u_n = 0$ and $R_n(t)r_n = Z(t)r_n$. Hence using (5.7) and (5.8) we get

$$| R_n(t)u | = | R_n(t)r_n | = | Z(t)r_n | \leq | Z(t) |_n | r_n |$$

$$\leq \text{const} | \exp [(\mu_{n+1} + \epsilon)t] | \; | u |$$

This proves the inequality (5.6).

If $Z(T)$ is compact, then by Corollary 4.2 the spectrum of B can have no purely imaginary values. The previous theorem therefore implies the following useful

Corollary 5.1. *If for some value of T the operator $Z(T)$ is compact, then $\{Z(t)\}$ decays exponentially.*

In case B has generalized eigenspaces the projection P_j is defined as

$$P_j = \frac{1}{2\pi i} \int_{\Gamma_j} (\mu I - B)^{-1} \, d\mu ,$$

where Γ_j is a small circle about the point μ_j containing no other point of $\sigma(B)$. It can be shown that the range of P_j is finite dimensional and that it reduces $Z(t)$. However, the action of $Z(t)$ on the range of P_j need no longer be multiplication by $\exp(\mu_j t)$ so that the jth term in (5.5) must be replaced by $Z(t)P_j$ which corresponds to a semigroup acting on a finite dimensional space having a generator whose spectrum consists of the single point μ_j. The proof of the more general statement is essentially the same as that of the simple case sketched above.

6.   Equivalent Incoming and Outgoing Representations

The scattering matrix is defined in terms of a pair of orthogonal incoming and outgoing subspaces D_- and D_+. To different pairs there correspond different scattering matrices. There are however pairs which are equivalent, in a sense to be made precise below; the aim of this section is to show that the scattering matrices corresponding to such equivalent pairs differ from each other by inessential factors.

Definition 6.1.   Two incoming (outgoing) subspaces D and D' are called *equivalent* with respect to the group $\{U(t)\}$ if there exists a real number T such that

(6.1) $U(T)D \subset D'$ and $U(T)D' \subset D.$

We also need the following concept:

Definition 6.2.   An operator-valued function $\mathfrak{M}(z)$ is called a *trivial inner factor* if

   (i) $\mathfrak{M}(z)$ is defined and unitary for z real;
   (ii) $\mathfrak{M}(z)$ can be extended as an analytic function to the lower half-plane and is there of exponential growth:

$$(6.2) \qquad\qquad | \, \mathfrak{M}(z) \, | \leq \exp\,(k \, | \, \mathrm{Im}\, z \, |);$$

   (iii) $\mathfrak{M}(z)$ is regular at every point of the lower half-plane and its inverse grows at most exponentially:

$$(6.3) \qquad\qquad | \, \mathfrak{M}^{-1}(z) \, | \leq \exp\,(k \, | \, \mathrm{Im}\, z \, |).$$

Remark 6.1.   Condition (iii) is equivalent to requiring that $\mathfrak{M}(z)$ be analytic in the upper half-plane and be of exponential growth there (see Lemma 1.3).

Remark 6.2.   The trivial inner factors form a group.

Theorem 6.1.   *Let D and D' be equivalent incoming [outgoing] subspaces. Then there exists a trivial inner factor $\mathfrak{M}(z)$ such that the spectral representatives a and a' of any element f in H with respect to the D and D' spectral representations, respectively, are related by*

$$(6.4) \qquad\qquad a'(\sigma) = \mathfrak{M}(\sigma)a(\sigma).$$

Proof.   Since a and a' are representatives of the same element f in two different incoming [outgoing] spectral representations, it follows that they are related by a unitary operator $\mathfrak{M}$ which commutes with $U(t)$. Also it follows from the assumed equivalence of D and D' that both $e^{iT\sigma}\mathfrak{M}(\sigma)$ and $e^{iT\sigma}\mathfrak{M}^{-1}(\sigma)$ map A_- $[A_+]$ into itself for some T. According to the theorem of Fourès and Segal (Theorem 4.1 of Chapter II), both operators can be realized as scattering matrices, that is as operator-valued functions unitary for real z analytic and bounded in the lower half-plane. Moreover, their product equals $\exp\,(2iT\sigma)I$ on the real axis and so is analytically extendable as $\exp\,(2iTz)I$; in particular their product equals $\exp\,(2iTz)I$ in the lower half-plane. This shows that $\mathfrak{M}(z)$ is a trivial inner factor and thus completes the proof of Theorem 6.1.

An immediate consequence is

Theorem 6.2. *If D_- , D_-' and D_+ , D_+' are pairs of equivalent incoming and outgoing subspaces, D_- orthogonal to D_+ and D_-' orthogonal to D_+', then the associated scattering matrices are related as follows:*

$$(6.5) \qquad\qquad S' = \mathfrak{M}_+ S \mathfrak{M}_-^{-1} \ ,$$

where $\mathfrak{M}_-$ and $\mathfrak{M}_+$ are the trivial inner factors relating a_- to a_-' and a_+ to a_+', respectively.

Since trivial inner factors are invertible for every z, formula (6.5) shows that the scattering matrices $S(z)$ and $S'(z)$ are regular at the same points in the lower half-plane; and that at points where they are singular the dimensions of the null spaces of $S(z)$ and $S'(z)$, as well as the codimensions of their ranges (that is the null spaces of $S^*(z)$ and $S'^*(z)$), are respectively equal. Since the location of points z where the scattering matrix is singular determines the spectrum of the infinitesimal generator of the related semigroup by Theorem 3.2, and since according to Corollary 3.2 the dimension of the null space of the scattering matrix equals the dimension of the corresponding eigenspace of the generator of the related semigroup, we have

Theorem 6.3. *Let D_- , D_-' and D_+ , D_+' be pairs of equivalent orthogonal incoming and outgoing subspaces and let $\{Z(t)\}$ and $\{Z'(t)\}$ be the related semigroups. Then the generators of $\{Z(t)\}$ and $\{Z'(t)\}$ have the same spectra and the corresponding eigenspaces are of the same dimension.*

Not only the eigenvalues but also the eigenvectors of equivalent semigroups are related. The following result will be used in Chapter V.

Theorem 6.4. *Let D_- , D_-' and D_+ , D_+' be pairs of equivalent incoming and outgoing orthogonal subspaces and let $\{Z(t)\}$ and $\{Z'(t)\}$ be the related semigroups. Assume in addition that D_- and D_+ contain D_-' and D_+', respectively. If μ is an eigenvalue of B, then the operator P_+ maps the nullspace of $\mu I - B'$ onto that of $\mu I - B$ in a one-to-one fashion.*

Proof. Since by Theorem 6.3 the null spaces of $\mu I - B$ and $\mu I - B'$ are of the same dimension, it suffices to show: If f' is an eigenvector of B', then

$$(6.6) \qquad\qquad f = P_+ f'$$

is a nonzero eigenvector of B corresponding to the same eigenvalue.

As a first step we show that f defined by (6.6) belongs to the domain K of $Z(t)$, that is that f is orthogonal to D_+ and D_-. By construction f is orthogonal to D_+; to prove that it is also orthogonal to D_- we make use of

Lemma 6.1. *f' is orthogonal to D_-.*

Proof of Lemma 6.1. Take any g in D_-. We wish to show that $(f', g) = 0$. Since f' is an eigenelement of B' we have

$$Z'(T)f' = e^{\mu T}f'.$$

Using this fact we obtain

$$(6.7) \qquad e^{\mu T}(f', g) = (e^{\mu T}f', g) = (Z'(T)f', g)$$

$$= (P_+'U(T)f', g) = (f', U(-T)P_+'g).$$

Now g belongs to D_- and is therefore orthogonal to D_+ and since D_+ contains D_+', g is *a fortiori* orthogonal to D_+'; thus $P_+'g = g$. On the other hand D_- is equivalent to D_-' and hence for T large enough $U(-T)$ maps D_- into D_-'. Thus it follows that for T large enough $U(-T)P_+'g$ belongs to D_-'. But since f' belongs to the domain K' of $Z'(t)$ it is orthogonal to D_-'. This shows that the scalar product on the extreme right in (6.7) is zero and hence proves the lemma.

Since f is defined as P_+f', we can write

$$(6.8) \qquad f = f' + h, \qquad h \quad \text{in} \quad D_+.$$

Now h is clearly orthogonal to D_- and by Lemma 6.1 so is f'; therefore, f itself is orthogonal to D_- and consequently lies in K.

To show that f is an eigenvector of B we proceed as follows: From (6.8) we have

$$Z(t)f = Z(t)f' + Z(t)h = Z(t)f'$$

since h is in D_+ and $Z(t)$ annihilates D_+. Now D_+ contains D_+' and hence $P_+ = P_+P_+'$, likewise $P_- = P_-'P_-$. Thus,

$$Z(t) = P_+U(t)P_- = P_+P_+'U(t)P_-'P_- = P_+Z'(t)P_-$$

and so

$$Z(t)f = Z(t)f' = P_+Z'(t)f' = P_+[e^{\mu t}f'] = e^{\mu t}f.$$

It remains only to show that f is not zero. We note that h in D_+ implies that $U(T)h$ lies in D_+' and hence that

$$Z'(T)h = P_+'U(T)h = 0 .$$

Thus $Z'(T)f = Z'(T)f' + Z'(T)h = \exp(\mu T)f' \neq 0$ and it follows that $f \neq 0$. This completes the proof of Theorem 6.4.

Given a pair of orthogonal incoming and outgoing subspaces D_- and D_+ one can construct a one parameter family of equivalent orthogonal pairs by setting

$$D_-' = D_-{}^a = U(-a)D_- \qquad \text{and} \qquad D_+' = D_+{}^a = U(a)D_+ ,$$

where a is any positive number. In this case the relation between the associated scattering matrices is particularly simple:

$$(6.9) \qquad\qquad S^a(z) = e^{-2iaz}S(z) .$$

Applying Theorem 6.4 to this case we have the following relation between the eigenvectors of $\mu I - B^a$ and those of $\mu I - B^b$ for $a < b$:

$$(6.10) \qquad\qquad f^a = P_+{}^a f^b .$$

We shall show in Chapter V how to construct projective limits of the eigenfunctions f^b by letting b tend to infinity. These limits turn out to be highly improper eigenfunctions of $\{U(t)\}$.

7. Notes and Remarks

Theorem 2.1 is the continuous analog of a series of results of Foias and Sz.-Nagy [1] on contraction operators. Foias and Sz.-Nagy consider the more complicated case where powers of the operator do not necessarily tend to zero strongly. In this case the associated "scattering matrix," called by them the characteristic operator function, need no longer be an inner factor in the sense of Beurling–Lax, that is it is not necessarily isometric on the boundary.

Theorem 3.2 relating the spectrum of the infinitesimal generator of the semigroup $Z(t)$ and the points where the scattering matrix is singular has been derived in the scalar case by Moeller [1]. The present extension to the operator case is an easy matter. Similar results have been derived by Foias and Sz.-Nagy [2] and by Helson in [1].

A more general version of the spectral mapping theorem than Theorem 4.1 is due to Phillips [1] and is presented in Hille–Phillips [1].

CHAPTER IV

The Translation Representation for the Solution of the Wave Equation in Free Space

Our ultimate aim is to apply the foregoing theory to the wave equation and thereby obtain a comparison of the asymptotic properties of the free space solution with those of the solution in an exterior domain. In the present chapter we shall develop the pertinent properties of the free space solution employing as our basic tool the translation representation.

As we shall see in Chapter V some of the properties of the exterior system can be deduced from the corresponding properties of the free space problem. For instance the incoming and outgoing subspaces are the same in both problems; but the fact that D_- is orthogonal to D_+ is most easily proved by means of the free space translation representation. Also explicit incoming and outgoing translation representations for the exterior problem will be obtained by means of the free space translation representation. Moreover, the eigenfunctions for the semigroup $\{Z(t)\}$ introduced in Chapter III will be characterized as μ-outgoing solutions of the reduced wave equations and these solutions are also studied in this chapter.

The development begins with the solution of the wave equation in free space for C_0^∞ data; this solution conserves energy and hence the class of solutions can be extended by continuity to all initial data of finite energy and the so-extended solution defines a group of unitary operators $\{U_0(t)\}$ with generator A_0. The subspaces D_- and D_+ of data for which the solutions vanish on $|x| < -t$ and $|x| < t$, respectively, are incoming and outgoing subspaces for $\{U_0(t)\}$. The spectral and translation representations for $\{U_0(t)\}$ are then constructed. The translation representation which is both incoming and outgoing turns out to be closely related to the Radon transform and has the additional property of being a unitary map of the initial data.

In order to study solutions of the reduced wave equation in this setup it is necessary to consider data of infinite energy and for this as well as the study of some properties of solutions with finite energy it is convenient to extend the above translation representation to distributions. The main difficulty in making this extension stems from the fact that the C_0^∞ data do not map onto all of the C_0^∞ translation representers. One consequence of this is that we are unable to obtain translation representers for all distribution-valued data and a second is that the dual mapping from distribution-valued representers to data is not one-to-one. However, it is possible to obtain the translation representer for distribution-valued data of bounded support as well as for what we call incoming and outgoing solutions of $(A_0 - \mu)f = g$, where g has compact support. Such solutions of $(A_0 - \mu)f = g$ give rise to μ-incoming and μ-outgoing solutions of the reduced wave equation $\Delta u - \mu^2 u = 0$ outside of the support of g. For Re $\mu \geq 0$ the resulting μ-outgoing solution satisfies the Sommerfeld radiation condition; however, the concept of μ-outgoing is applicable for all complex μ and as noted above it plays an essential role in our discussion of the eigenfunctions of $\{Z(t)\}$.

All of our considerations are limited to the case of an odd number of spatial dimensions.

1.   The Hilbert Space H_0 and the Group $\{U_0(t)\}$

The elements of H_0 are the Cauchy data for the wave equation, that is pairs of complex valued functions defined over the entire space R_n :

$$f = \{ f_1, f_2 \} .$$

Taking f_1 and f_2 to be smooth with compact support we define the *energy norm* of f as

(1.1)
$$| f |_E^2 = \frac{1}{2} \int \{| \partial_x f_1 |^2 + | f_2 |^2\} \, dx ;$$

integration is over R_n , and $| \partial_x f_1 |^2$ stands for $\sum_{j=1}^n | \partial f_1 / \partial x_j |^2$. The corresponding scalar product will be denoted as $(f, g)_E$.

H_0 is defined as the completion of the above space of C_0^∞ pairs f in the energy norm. Consequently, the elements of H_0 can be represented as pairs of vectors of which the second is a square integrable function and the first belongs to the completion of C_0^∞ in the norm $\{\int |\partial_x f_1|^2\}^{1/2}$. It is useful to know that

Lemma 1.1. *If the number n of variables exceeds 2, then the first component of an element of H_0 is a locally square integrable function.*

The proof of this is based on the following *a priori* inequality for functions f_1 in C_0^∞:

$$(1.2) \qquad \int_{|x|\leq R} |f_1|^2\, dx \leq \frac{R^2}{2(n-2)} \int |\partial_x f_1|^2\, dx\,,$$

which can be derived as follows: Write

$$f_1(x) = -\int_{|x|}^\infty \partial_r f_1\left(r\,\frac{x}{|x|}\right) dr\,.$$

By the Schwarz inequality

$$|f_1(x)|^2 \leq \left(\int_{|x|}^\infty r^{1-n}\, dr\right)\left(\int_{|x|}^\infty |\partial_r f_1|^2\, r^{n-1}\, dr\right)$$

$$= \frac{|x|^{2-n}}{n-2}\int_{|x|}^\infty |\partial_r f_1|^2\, r^{n-1}\, dr\,.$$

Integrating with respect to $\omega = x/|x|$ on the unit sphere gives us

$$\int |f_1(R\omega)|^2\, d\omega \leq \frac{R^{2-n}}{n-2}\int_{|x|\geq R} |\partial_x f_1|^2\, dx\,;$$

(1.2) follows from this after multiplication by R^{n-1} and an integration.

Let $u(x, t)$ be the free space solution of the wave equation with initial value f:

$$u_{tt} - \Delta u = 0$$

$$(1.3) \qquad u(x, 0) = f_1(x)\,, \qquad u_t(x, 0) = f_2(x)\,.$$

The following result is classical:

Theorem 1.1. *Given data f of class C_0^∞, the initial value problem (1.3) has a unique C^∞ solution u whose energy does not vary with t.*

Proof.   The construction of a solution is easily accomplished by means of the Fourier transformation (see Section 2 of this chapter or any text on the theory of partial differential equations). To prove the conservation of energy we multiply the wave equation (1.3) by u_t and integrate over the slab $0 \leq t \leq T$. Integration by parts gives

$$0 = \iint u_t(u_{tt} - u_{xx}) \, dx \, dt = \iint (u_t u_{tt} + u_{tx} u_x) \, dx \, dt$$

$$(1.4) \qquad = \iint \tfrac{1}{2} (u_t^2 + u_x^2)_t \, dx \, dt = \frac{1}{2} \int (u_t^2 + u_x^2) \, dx \bigg|_0^T,$$

as asserted. The second equality in (1.4) is a consequence of the finite velocity of the signal which follows from

Theorem 1.2.   *Suppose that the data f are zero in the ball $|x - x_0| < R$; then the solution $u(x, T)$ of the initial value problem (1.3) is zero in the ball $|x - x_0| < R - T$.*

The proof of this goes the same way as that of the conservation of energy, except that this time we perform the integration (1.4) over the truncated cone $|x - x_0| < R - t$, $0 \leq t \leq T$, and note that the terms on the lateral boundaries $0 < t < T$ are positive.

Theorem 1.2 asserts that *no signal is transmitted with speed greater than one.* The next result asserts that *if the number of space variables is odd, then no signals are transmitted with speed less than one:*

Theorem 1.3.   *Suppose that the data f are zero for $|x - x_0| > R$, then the solution $u(x, t)$ of the initial value problem is zero for $|x - x_0| < |t| - R$.*

A proof of this will be given in Section 2. Combining the last two results we obtain the celebrated

Huygens' Principle.   *The value of a solution $u(x, t)$ of the wave equation at the point (x, t) depends only on values of its initial data and of their derivatives at points of the sphere $|x - x_0| = t$*

We now *define* the operator $U_0(t)$ as mapping initial data of solutions of the wave equation into the solution data at time t·

$$U_0(t): \quad \{f_1, f_2\} \rightarrow \{u(t), u_t(t)\}.$$

It follows from Theorems 1.1 and 1.2 that the operator $U_0(t)$ as defined above maps C_0^∞ data into C_0^∞ data, forms a one parameter group and conserves energy. Therefore it can be extended by continuity to all of H_0, and defines a *one-parameter group of unitary operators on H_0*.

According to the theorem of Stone every one parameter unitary group has an infinitesimal generator which is skew-selfadjoint. *Let A_0 denote the infinitesimal generator of $\{U_0(t)\}$. Clearly every f in C_0^∞ belongs to the domain of A_0, and for such f*

$$A_0 f = \frac{d}{dt} U_0(t) f \big|_{t=0} = \begin{pmatrix} u_t \\ u_{tt} \end{pmatrix}\bigg|_{t=0} = \begin{pmatrix} u_t \\ \Delta u \end{pmatrix}\bigg|_{t=0} = \begin{pmatrix} f_2 \\ \Delta f_1 \end{pmatrix} = \begin{pmatrix} 0 & I \\ \Delta & 0 \end{pmatrix} f.$$

More generally one can show (see Section 2) that *A_0 is the closure of matrix differential operator*

(1.5)
$$\begin{pmatrix} 0 & 1 \\ \Delta & 0 \end{pmatrix}$$

defined originally in C_0^∞. Since, as shown in Theorem 1.2, signals propagate with finite speed, the initial value problem (1.3) can be solved for arbitrary C^∞ data f, subject to no restriction whatsoever at infinity. An even further extension is possible by admitting distributions as initial data; of course in this case the solutions are also distributions:

Theorem 1.4. *Given a pair of distributions f_1 and f_2 there exists a unique distribution solution $u(t)$ of the wave equation with initial data f in the sense that for every C_0^∞ function $\phi(x)$, (u, ϕ) is a C^∞ function of t, equal to (f_1, ϕ) at $t = 0$ and whose derivative at $t = 0$ equals (f_2, ϕ).*

Proof. Theorem 1.4 is the dual to the existence Theorem 1.1 for C_0^∞ data; however, the correspondence is somewhat complicated. Suppose first that f, g belong to H_0; then from the unitary property of $\{U_0(t)\}$ we see

$$(U_0(t)f, g)_E = (f, U_0(-t)g)_E$$

and for g with $A_0 g$ in C_0^∞ this can be written as

$$([U_0(t)f]_1, -\Delta g_1) + ([U_0(t)f]_2, g_2)$$
$$= (f_1, -\Delta[U_0(-t)g]_1) + (f_2, [U_0(-t)g]_2)$$

where $(\cdot, \cdot)$ stands for the usual $L_2(R_n)$ inner product. Since g lies in the domain of A_0, $A_0 U_0(-t)g = U_0(-t)A_0 g$; that is

$$U_0(-t)\{g_2, \Delta g_1\} = \{[U_0(-t)g]_2, \Delta[U_0(-t)g]_1\} .$$

Thus the above expression can be rewritten as

$$([U_0(t)f]_1, -\Delta g_1) + ([U_0(t)f]_2, g_2)$$
$$= (f_1, -[U_0(-t)\{g_2, \Delta g_1\}]_2) + (f_2, [U_0(-t)\{g_2, \Delta g_1\}]_1) .$$

Finally, we note that for any $\{\phi, \psi\}$ in C_0^∞ the potential solution g_1 of $\Delta g_1 = -\phi$, is such that $\{g_1, \psi\}$ lies in the domain of A_0. We can therefore define the distribution extensions $U_0(t)f$ by

$$([U_0(t)f]_1, \phi) = (f_1, [U_0(-t)\{0, \phi\}]_2) - (f_2, [U_0(-t)\{0, \phi\}]_1) ,$$

(1.6)

$$([U_0(t)f]_2, \psi) = -(f_1, [U_0(-t)\{\psi, 0\}]_2) + (f_2, [U_0(-t)\{\psi, 0\}]_1) .$$

It is readily verified that $u = [U_0(t)f]_1$, as defined in (1.6), is the desired distribution solution of the wave equation.

It remains to establish the uniqueness and for this we may suppose that u is a distribution solution with vanishing initial data. Let $\phi(x, t)$ be an arbitrary C^∞ solution of the wave equation with compact support for each t and set

$$l(t) = (u, \phi_t) - (u_t, \phi) .$$

Then

$$\frac{dl}{dt} = (u, \phi_{tt}) - (u_{tt}, \phi) = (u, \Delta\phi) - (u, \Delta\phi) = 0.$$

Since $l(0) = 0$ we see that $l(t)$ is always zero. For any t_0 and any ψ in C_0^∞ we may choose $\phi(x, t)$ so that $\phi(x, t_0) = 0$ and $\phi_t(x, t_0) = \psi$. Thus, $(u(t_0), \psi) = 0$ for all ψ in C_0^∞ and hence $u(t_0) = 0$.

The group of operators $\{U_0(t)\}$ can then be extended to distributions as well. We shall denote by W the operator assigning to data f the solution u of the wave equation with initial data f.

There is one more theorem about the wave equation in free space which will be needed in Chapter V:

Theorem 1.5 (Holmgren Uniqueness Theorem). *Let $u(x, t)$ be a solution of the wave equation which is zero inside the cylinder $|x| < R$, $|t| \leq T$. Then $u(x, t)$ is zero for $|x| < R + T - |t|$.*

A proof of this based on Corollary 3.1 will be given in Section 3.

2. Spectral and Translation Representations of $\{U_0(t)\}$

Let $D_+[D_-]$ respectively denote the set of data f in H_0 which have the property that the corresponding solutions $u = Wf$ of the wave equation vanish in the forward [backward] cone:

$$(2.1) \qquad |x| < t \qquad [|x| < -t].$$

We claim that D_+ is outgoing in the sense of Chapter II, i.e., that D_+ is closed and satisfies the conditions

(i) $U_0(t)D_+ \subset D_+$ for $t > 0,$

(ii) $\cap\, U_0(t)D_+ = \{0\},$

(iii) $\overline{\cup\, U_0(t)D_+} = H_0\,;$

and that D_- satisfies the analogous incoming conditions.

That D_+ is closed follows from the estimate obtained in the proof of Lemma 1.1. Property (i) is clearly satisfied since if $u(x, t) = Wf$ vanishes in the cone (2.1) then for $s > 0$, $WU_0(s)f = u(x, t + s)$ vanishes in the larger cone: $|x| < t + s$. This shows that all data in $U_0(s)D_+$ are zero for $|x| < s$; and (ii) follows from this. To verify (iii) we use Huygens' principle to conclude that if f is zero for $|x| \leq R$ then $u(x, t) = Wf$ is zero for $|x| < t - R$; this implies that $U_0(R)f = g$ belongs to D_+ and hence $U_0(-R)D_+$ includes all data with support in the ball $|x| \leq R$. Since H_0 is the completion of data with compact support, (iii) follows.

We can therefore apply the results of Chapter II and conclude that there exist incoming and outgoing translation representations for the group $\{U_0(t)\}$. We shall show furthermore in this and the next section that D_+ and D_- are orthogonal complements of each other:

$$D_+ \oplus D_- = H_0\,;$$

from this it would follow that the incoming and outgoing representations are one and the same.

The steps of the actual proof will be carried out in reverse order. We shall construct a particular translation representation for $\{U_0(t)\}$ and then we shall verify that the incoming and outgoing subspaces associated with this representation are D_+ and D_- as defined above. The above translation representation will in turn be obtained by Fourier transformation from a spectral representation for $\{U_0(t)\}$.

In accordance with general spectral theory the spectral representation of a given f in H_0 ought to be given as a scalar product with eigenfunctions or generalized eigenfunctions ϕ_σ of the infinitesimal generator A_0 of $\{U_0(t)\}$:

$$\tilde{f}(\sigma) = (f, \phi_\sigma)_E$$

where ϕ_σ satisfies the eigenvalue equation

(2.2) $$A_0\phi_\sigma = i\sigma\phi_\sigma .$$

In Section 1 we have determined the form of A_0 :

$$A_0 = \begin{pmatrix} 0 & 1 \\ \Delta & 0 \end{pmatrix}.$$

This operator has no proper (square integrable) eigenfunctions; however, among its improper but bounded eigenfunctions are the functions

(2.3) $$\phi_{\sigma,\omega} = \exp(-i\sigma x\omega)\{1, i\sigma\} \qquad (|\omega| = 1).$$

There are infinitely many linearly independent bounded eigenfunctions for each σ, indicating that the spectral multiplicity of A_0 is infinite. Accordingly $\tilde{f}$ will be a function of both σ and ω; equivalently we can regard $\tilde{f}$ as function of σ whose values lie in the auxiliary space $N = L_2(S_{n-1})$. We shall denote the L_2 scalar product over $R \times S_{n-1}$ by square brackets:

$$[\tilde{f}, \tilde{g}] = \iint \tilde{f}(s, \omega)\bar{\tilde{g}}(s, \omega) \, ds \, d\omega.$$

In order to make the spectral representation unitary, the eigenfunctions have to be properly weighted. An easy calculation, carried out below, shows that the proper weight factor is of the form $\sigma^{(n-3)/2}$:

Theorem 2.1. *The function $\tilde{f}$ defined by*

(2.4) $$\tilde{f}(\sigma, \omega) = (-i\sigma)^{(n-3)/2} \frac{1}{(2\pi)^{n/2}} (f, \phi_{\sigma,\omega})_E ,$$

where $\phi_{\sigma,\omega}$ is given by (2.3), *is a unitary spectral representation of H_0 for* $\{U_0(t)\}$.

Before carrying out the verification we shall state as a further theorem the explicit expression for the corresponding translation representation:

Theorem 2.2. *Let k be the translation representation of f in S derived from the spectral representation (2.4) by Fourier transformation; then k and f are expressed in terms of each other by the following formulae:*

$$(2.5) \qquad k(s, \omega) = -\partial_s{}^{(n+1)/2} M_1 + \partial_s{}^{(n-1)/2} M_2$$

where M_j, $j = 1, 2$ are defined as the following integrals over hyperplanes:

$$(2.6) \qquad M_j(s, \omega) = \frac{1}{2} \frac{1}{(2\pi)^{(n-1)/2}} \int_{x\omega = s} f_j(x) \, dS.$$

Conversely,

$$(2.7) \qquad f_1(x) = S(x), \qquad f_2(x) = S'(x)$$

where S, S' are defined as the following integrals over spheres:

$$(2.8) \qquad S(x) = \int h(x\omega, \omega) \, d\omega, \qquad S'(x) = \int h'(x\omega, \omega) \, d\omega.$$

Here, h and h' are abbreviations for the following functions:

$$(2.9) \qquad \begin{aligned} h(s, \omega) &= \frac{1}{(2\pi)^{(n-1)/2}} (-\partial_s)^{(n-3)/2} k, \\[2ex] h'(s, \omega) &= \frac{1}{(2\pi)^{(n-1)/2}} (-\partial_s)^{(n-1)/2} k. \end{aligned}$$

Proof of Theorem 2.1. Take f to be of class S, i.e., C^∞ and such that all derivatives of f tend to zero at infinity faster than any power of $|x|^{-1}$. It follows from simple estimates that $\tilde{f}$ as defined by (2.4) also belongs to class S. Denote by $\tilde{h}$ the representer of $A_0 f$; we get by integration by parts and the fact that ϕ is an eigenfunction of A_0 that

$$\tilde{h} = (A_0 f, \phi)_E = -(f, A_0\phi)_E = i\sigma(f, \phi)_E = i\sigma\tilde{f}.$$

This shows that (2.4) is indeed a spectral representation for A_0. From this it follows easily that (2.4) also defines a spectral representation for $\{U_0(t)\}$.

To prove that the representation is isometric we perform an integration by parts which converts the energy scalar product into an L_2 scalar product;

$$(2.10) \quad (f, \phi)_E = \frac{1}{2} \int (\partial_x f_1 \, \overline{\partial_x \phi_1} + f_2 \bar{\phi}_2) \, dx = \frac{1}{2} \int (-f_1 \, \overline{\Delta \phi_1} + f_2 \bar{\phi}_2) \, dx .$$

Using the definition (2.3) of $\phi_{\sigma, \omega}$ we see that

$$(2.11) \qquad \tilde{f} = -(-i\sigma)^{(n+1)/2} \tilde{f}_1 + (-i\sigma)^{(n-1)/2} \tilde{f}_2$$

where

$$(2.12) \qquad \tilde{f}_j(\sigma, \omega) = \frac{1}{2} \frac{1}{(2\pi)^{n/2}} \int f_j(x) e^{i\sigma x \omega} \, dx \qquad (j = 1, 2) .$$

Formula (2.12) shows that both functions $\tilde{f}_j$ are even functions of (σ, ω); it follows from this that the functions on the right in (2.11) have opposite parity and are thus orthogonal:

$$(2.13) \qquad || \tilde{f} ||^2 = || \sigma^{(n+1)/2} \tilde{f}_1 ||^2 + || \sigma^{(n-1)/2} \tilde{f}_2 ||^2 .$$

Since the functions $\tilde{f}_j$ are even,

$$|| \sigma^{(n+1)/2} \tilde{f}_1 ||^2 = 2 \int_0^\infty | \tilde{f}_1(\sigma, \omega) |^2 \sigma^{n+1} \, d\sigma \, d\omega$$

$$(2.14)$$

$$|| \sigma^{(n-1)/2} \tilde{f}_2 ||^2 = 2 \int_0^\infty | \tilde{f}_2(\sigma, \omega) |^2 \sigma^{n-1} \, d\sigma \, d\omega .$$

Formulas (2.12) show that $\tilde{f}_j$ can be expressed in terms of the Fourier transform $\hat{f}_j(\xi)$ of f_j by setting $\xi = \sigma \omega$; using this and the Parseval formula in (2.14) we get

$$|| \sigma^{(n+1)/2} \tilde{f}_1 ||^2 = \frac{1}{2} \int | \partial_x f_1 |^2 \, dx$$

$$|| \sigma^{(n-1)/2} \tilde{f}_2 ||^2 = \frac{1}{2} \int | f_2 |^2 \, dx$$

which, when substituted into (2.13), gives the isometry:

$$|| \tilde{f} ||^2 = | f |_E^2 .$$

To show that the representation is not only isometric but unitary we have to verify that the set of function $\tilde{f}$ which represent data f of class S are dense in $L_2(R \times S_{n-1})$. It follows from formulas (2.11) and (2.12) and from the fact that the two terms on the right in (2.11) are of opposite parity, that if $\tilde{f}$ is C^∞ and vanishes for σ near zero and near infinity then $\hat{f}_1$ and $\hat{f}_2$ are C^∞ functions with compact support and so f_1 and f_2 belong to class S. Since the above class of functions $\tilde{f}$ is dense in $L_2(R \times S_{n-1})$, our proof of Theorem 2.1 is complete.

As we have already noted A_0 is simply multiplication by $i\sigma$ in the spectral representation where the domain of A_0 is precisely the set of all $\tilde{f}$ in $L_2(R \times S_{n-1})$ such that $\sigma \tilde{f}$ also belongs to $L_2(R \times S_{n-1})$. Function pairs $\{\tilde{f}, i\sigma \tilde{f}\}$ with compact support are obviously dense in the graph of A_0. Thus an argument similar to the above can be used to show that A_0 is the closure of its restriction to the set of functions which represent data in S. Let $\phi(s)$ in C_0^∞ be identically 1 for small $|s|$, then for f in S

$$\{\phi(|x|/n)f, A_0\phi(|x|/n)f\} \rightarrow \{f, A_0 f\}$$

in the topology of $H_0 \times H_0$. This proves that A_0 is also the closure of its restriction to data of class C_0^∞, as asserted in Section 1.

We turn now to the proof of Theorem 2.2. Assume that f belongs to class S; we start by performing the integrations in (2.12) first along the hyperplanes $x\omega = s$ and then with respect to s; we get

$$(2.15) \qquad \tilde{f}_j(\sigma, \omega) = \frac{1}{(2\pi)^{1/2}} \int M_j(s, \omega) e^{i\sigma s} \, ds \qquad (j = 1, 2),$$

where M_j is given by formula (2.6).

Substituting (2.15) into (2.11) gives

$$\tilde{f}(\sigma, \omega) = \frac{1}{(2\pi)^{1/2}} \int [-M_1(s, \omega)(-i\sigma)^{(n+1)/2} + M_2(s, \omega)(-i\sigma)^{(n-1)/2}] e^{i\sigma s} \, ds.$$

An integration by parts now shown that $\tilde{f}$ is the Fourier transform of the function given in formula (2.5); this completes the proof of the first part of Theorem 2.2.

To prove the second half we have to invert (2.5); for this we shall use the unitary character of the translation representation, according to which

$$(2.16) \qquad (f, g)_E = [k, l]$$

for all f and g in H_0, k and l being the translation representations of f and g, respectively. It is enough to assume that g belongs to class S and that

k is a C^∞ function of s and ω in $L_2(R \times S_{n-1})$. Using formula (2.5) to express l on the right in (2.16) we get

$$[k, l] = \iint k(s, \omega)\,(-\partial_s^{(n+1)/2}\bar{M}_1 + \partial_s^{(n-1)/2}\bar{M}_2)\,ds\,d\omega\,.$$

Integrating by parts with respect to s gives

$$[k, l] = (2\pi)^{(n-1)/2} \iint (h_1\bar{M}_1 + h_2\bar{M}_2)\,ds\,d\omega$$

where

$$(2.17) \quad h_1 = \frac{-1}{(2\pi)^{(n-1)/2}}\,(-\partial_s)^{(n+1)/2}k\,, \qquad h_2 = \frac{1}{(2\pi)^{(n-1)/2}}\,(-\partial_s)^{(n-1)/2}k\,.$$

Substitute the explicit expression (2.6) for M_1 and M_2 in the above and recombine dS and ds again as dx in the resulting multiple integral; we then have

$$[k, l] = \frac{1}{2} \iint h_1 \left(\int_{x\omega=s} \bar{g}_1(x)\,dS \right) ds\,d\omega + \iint h_2 \left(\int_{x\omega=s} \bar{g}_2(x)\,dS \right) ds\,d\omega$$

$$= \frac{1}{2} \iint (h_1(x\omega, \omega)\bar{g}_1(x) + h_2(x\omega, \omega)\bar{g}_2(x))\,dx\,d\omega\,.$$

An interchange in the order of integration with respect to x and ω now gives

$$(2.18) \qquad [k, l] = \frac{1}{2} \int (S_1(x)\bar{g}_1(x)\,dx + S_2(x)\bar{g}_2(x))\,dx$$

where

$$(2.19) \qquad S_j(x) = \int h_j(x\omega, \omega)\,d\omega\,.$$

On the other hand using (2.10), we can write the left side of (2.16) as

$$(2.18)' \qquad (f, g)_E = \frac{1}{2} \int (-\Delta f_1\,\bar{g}_1 + f_2\bar{g}_2)\,dx\,.$$

In view of (2.16), (2.18) and (2.18)$'$ must be equal for all g of class S. Since such data form a dense set, the coefficients of g_1 and g_2 in (2.18) and

$(2.18)'$ are identical:

$$(2.20) \qquad -\Delta f_1 = S_1 , \qquad f_2 = S_2 ;$$

this proves the second of the two formulas given for f in (2.7) of Theorem 2.2. To obtain the first formula we express the function S_1 as given by (2.19) in terms of S given by (2.8):

$$S_1 = -\Delta S .$$

Combining this with the first equation of (2.20) gives

$$\Delta(f_1 - S) = 0 ,$$

i.e., $f_1 - S$ is a harmonic function defined in the whole space. On the other hand if f is of class s then so is h and hence (2.8) show that S tends to zero as $|x| \to \infty$. Thus $f_1 - S$ tends to zero as $|x| \to \infty$ and therefore, by the maximum principle, it is identically zero. So $f_1 = S$, as asserted in (2.7).

Applying (2.7) to $U_0(t)f$, and noting that the latter is represented by the translate of k we obtain

Corollary 2.1. *Let* $u(x, t) = Wf$ *be the solution of the wave equation with initial data* f *and assume that the translation representer of* f *is* C^∞. *Then*

$$(2.21) \qquad u(x, t) = \int h(x\omega - t, \omega)\, d\omega ,$$

$$(2.21)' \qquad u_t(x, t) = \int h'(x\omega - t, \omega)\, d\omega .$$

As we shall see in Section 3, formula (2.21) holds in a somewhat weaker sense for a very general class of data.

Suppose that the data f are C^∞ and equal to zero for $|x| \geq R$; then it follows from (2.5) and (2.6) that k is zero for $|s| \geq R$. From this it follows by (2.9) and (2.21) that $u(x, t) = Wf$ is zero inside the cones $|x| < |t| - R$, for then $|x\omega - t| > R$ and hence the integrand in (2.21) is zero for all ω in S_{n-1}. This furnishes a proof for Huygens' principle, stated as Theorem 1.3 in Section 1.

Theorem 2.3. *The subspaces* $L_2((-\infty, 0) \times S_{n-1})$ *and* $L_2((0, \infty) \times S_{n-1})$ *associated with the translation representation determined in Theorem 2.2 are the spaces* D_- *and* D_+ *respectively described at the beginning of this section.*

Proof. Suppose that f in S is represented by a function k which vanishes on the negative axis. Then formula (2.9) shows that h also vanishes there and it follows from formula (2.21) that $u(x, t)$ is zero in the forward cone (2.1). More generally take any f in H_0 which is represented by a function k vanishing on the negative axis. Approximate k by a sequence of C^∞ functions k_n ; which vanish on the negative axis; it follows from (2.21) that the elements f_n represented by k_n belong to D_+ ; and since D_+ is closed it follows that f itself belongs to D_+ .

The proof of the converse—that every f in D_+ is represented by a function k which vanishes on the negative axis—is more difficult and will be proved in Section 3 from Corollary 3.2.

Remark 1.   Signals with initial data in D_+ "go out" to infinity in the sense that at time t the signal is zero inside the ball $|x| < t$; this is the origin of the phrase "outgoing," and similarly for "incoming."

Remark 2.   In constructing the spectral representation (2.4) the absolute value of the weight factor, $(\sigma)^{(n-3)/2}$, was dictated by reasons of isometry, but its argument was entirely arbitrary. Certainly taking the argument to be constant for all σ, ω is the simplest choice but it is somewhat fortuitous that this "simplest" choice is the one that gives D_+ and D_- as $L_2((0, \infty) \times S_{n-1})$ and $L_2((-\infty, 0) \times S_{n-1})$, respectively.

An immediate consequence of Theorem 2.3 is

Corollary 2.2.   *D_+ and D_- are orthogonal.*

Suppose that $f = \{f_1, f_2\}$ belongs to D_+ ; then $f' = \{f_1, -f_2\}$ belongs to D_- and so by Corollary 2.2 f and f' are orthogonal:

$$0 = (f, f')_E = \frac{1}{2} \left\{ \int |\partial_x f_1|^2 \, dx - \int |f_2|^2 \, dx \right\}.$$

Making use of the physical interpretation of the two energy terms, we can state the above identity as

Corollary 2.3.   *In an outgoing or incoming signal the kinetic and potential energies are equipartitioned.*

Another useful consequence of Theorem 2.3 is

Corollary 2.4. *The span of*

$$F = [U_0(t)f \, ; \, \mathrm{supp}\, f \subset \{|\, x\,| < t\}]$$

is dense in D_+ .

Proof. If the corollary were not true then there would exist a nontrivial g in D_+ orthogonal to F. In this case

$$(U_0(-t)g, f)_E = (g, U_0(t)f)_E = 0$$

for all f with support in the ball $\{|\, x\,| < t\}$. Setting $u = Wg$ we see that $u(x, t)$ is harmonic and $u_t(x, t) = 0$ in the ball $\{|\, x\,| < -t\}$. Thus there is a function ϕ harmonic in all R_n such that $u(x, t) = \phi(x)$ for $|\, x\,| < -t$. Moreover, $|\, \partial_x \phi\,|^2$ is integrable and this requires that ϕ be constant. Hence, $u(x, t) = c$ for $|\, x\,| < -t$ and applying the inequality (1.2) with $R = -t$ and $f_1 = u(x, t)$ we obtain

$$c^2 \,|\, t\,|^{n-2} \le \mathrm{const}\, |\, g\,|_E^2$$

for all $t < 0$. Consequently, $c = 0$ and g belongs to both D_- and D_+ ; but according to Corollary 2.2 this implies that $g = 0$, contrary to our choice of g.

We shall now show how formula (2.21) can be used to study the asymptotic behavior of solutions of the wave equation along rays, i.e., along lines of the form $x = (t + s)\theta$, as t tends to $\pm \infty$. Setting $x = (t + s)\theta$ in formula (2.21)$'$ we obtain

$$(2.22) \qquad u_t((t + s)\theta, t) = \int h'[(t + s)(\theta\omega - 1) + s, \omega]\, d\omega.$$

Assume that h' is continuous and has a bounded support. Then for $|\, t\,|$ large the integrand on the right in (2.22) is zero except for values of ω near θ. So in replacing ω by θ in the second argument we introduce an error ϵ which tends to zero as $|\, t\,| \to \infty$. Making this replacement and introducing $\rho = \theta\omega$ as the new variable of integration leads to the following expression for the right side of (2.22):

$$\omega_{n-1} \int_{-1}^{1} \{h'[(t + s)(\rho - 1) + s, \theta] + \epsilon\} (1 - \rho^2)^{(n-3)/2} \, d\rho,$$

where ω_{n-1} is the area of the unit sphere in R_{n-1} . Switching to $\tau = |\, t + s\,| \times$

$(1 - \rho)$ as new variable of integration transforms this into

$$2^{(n-3)/2}\omega_{n-1} \mid t + s \mid^{(1-n)/2} \int_0^\infty \{h'(s \mp \tau, \theta) + \epsilon\}\tau^{(n-3)/2}(1 + \eta) \, d\tau,$$

where η is a polynomial in τ which tends to zero as $\mid t + s \mid \to \infty$.

According to formula (2.9), $h' = (2\pi)^{(1-n)/2}(-\partial_s)^{(n-1)/2}k$; substituting this into the above integral, integrating by parts $(n - 1)/2$ times, and replacing ω_{n-1} by $2(\pi)^{(n-1)/2}[((n - 3)/2)!]^{-1}$ gives

$$(t + s)^{(n-1)/2}u_t((t + s)\theta, t) = k(s, \theta) + o(1).$$

So we have proved

Theorem 2.4. *Let $u(x, t)$ be a solution of the wave equation with finite energy, and let k denote the translation representation of the initial values of u constructed in Theorem 2.2. Assume that $\partial_s^{(n-1)/2}k$ is continuous and has bounded support. Then*

(2.23) $$\lim_{|t| \to \infty} t^{(n-1)/2}u_t(x, t) = k(s, \theta)$$

along the ray

(2.24) $$x = (t + s)\theta.$$

As we shall see in Chapter V an analogous result holds for wave propagation in the presence of obstacles. In this case the limit (2.23) in the *positive* t direction gives the value of the *outgoing* translation representation of the initial data of u, and the limit in the *negative* t direction gives the value of the *incoming* translation representation of these data. The scattering operator therefore relates the behavior of solutions along rays for large positive and negative times.

We shall denote by $\Re$ the operator assigning the representer k defined by formulas (2.5), (2.6) to data f:

$$k = \Re f.$$

The operator relating f to k, defined by formulas (2.7), (2.8), (2.9) will be denoted by $\mathcal{G}$:

$$f = \mathcal{G}k.$$

For the classes of data f and representers k considered in this section these operators are inverse to each other: $\mathfrak{R}\mathcal{J} = \mathcal{J}\mathfrak{R} = I$. In Sections 3 and 4 we shall extend these operators; the extended $\mathcal{J}$ will still be a left inverse to the extended $\mathfrak{R}$ but it will not be a right inverse.

3. The Operator $\mathcal{J}$ Extended to Distributions

In this section we shall be dealing with initial data $f = \{f_1, f_2\}$ where f_1 and f_2 are distributions. As remarked at the end of Section 1, the initial value problem for the wave equation has a unique solution u for arbitrarily given distribution data; that is $U_0(t)f$ and $u = Wf$ are well defined as distributions. Similarly, $A_0 f$ is well defined, where A_0 denotes the infinitesimal generator of $\{U_0(t)\}$.

Given a pair of data f and g, one a distribution and the other C^∞, whose supports have compact intersection, then the energy scalar product

$$(f, g)_E = \frac{1}{2}\int \partial_x f_1 \, \overline{\partial_x g_1} + f_2 \bar{g}_2 \, dx$$

is well defined. The following result follows immediately by integration by parts:

Lemma 3.1. *Let G be an open set, f a given distribution-valued Cauchy data. Then*

$$(g, f)_E = 0$$

for all C^∞ data g with compact support contained in G if and only if $A_0 f = 0$ in G.

We recall from the definition of A_0 that $A_0 f = 0$ in G means that f_1 is harmonic in G and f_2 is zero there.

We turn now to the task of defining the operator $\mathcal{J}$ for distributions. This is easily accomplished since formulae (2.7), (2.8), and (2.9) can be used to define $\mathcal{J}k$ for any C_0^∞ function k. It is furthermore easy to show that $\mathcal{J}$ thus defined is continuous in the weak sequential topology for distributions; that is, if a sequence $\{k_n\}$ of C_0^∞ functions converges to a distribution k in the sense that

$$\lim [\phi, k_n] = [\phi, k]$$

for every C_0^∞ test function ϕ, then $\mathcal{J}k_n$ tends weakly to a distribution f. This follows from C_0^∞ data g having C_0^∞ translation representers and the fact that if we define M_1 and M_2 as in (2.6) for g rather than f, then†

$$(f_1, g_1) = 2[k, \partial_s^{(n-3)/2}M_1],$$

$$(f_2, g_2) = 2[k, \partial_s^{(n-1)/2}M_2].$$

This weak limit f is defined to be $\mathcal{G}k$. We shall call k a translation representation of $\mathcal{G}k$. In Section 2 we extended the operator $\mathcal{G}$ by continuity to all L_2 functions k. Obviously, this further extension is consistent with the previous one.

The extended operator $\mathcal{G}$ retains most of its former properties; we list them below for convenient reference. All of them can be deduced by passage to the limit from the C^∞ case:

Properties of the operator $\mathcal{G}$:

(3.1a) $\partial_x \mathcal{G} = \mathcal{G}\omega \, \partial_s$

(3.1b) $A_0 \mathcal{G} = -\mathcal{G} \, \partial_s$

(3.1c) $U_0(t)\mathcal{G} = \mathcal{G}T(t)$

where $T(t)$ denotes right translation by amount t,

(3.1d) $O\mathcal{G} = \mathcal{G}O$

where O denotes any orthogonal transformation acting on the x variables in the left member of (3.1d) and on ω in the right member. If

(3.1e) $k = 0$ for $|s| < r,$ then $\mathcal{G}k = 0$ for $|x| < r.$

$\mathcal{G}$ is adjoint to $\mathcal{R}$ in the sense that

(3.1f) $(g, \mathcal{G}k)_E = [\mathcal{R}g, k]$

for all distributions k and all C_0^∞ data g.

$u = W\mathcal{G}k$ satisfies the following weak form of (2.21): For any C_0^∞ func-

† For f in H_0 and $A_0^2\{h, 0\} = \{g_1, 0\}$

$$(f_1, g_1) = -\tfrac{1}{2} (f, \{h, 0\})_E = 2 [k, \partial_s^{(n+1)/2}M_1'] = 2[k, \partial_s^{(n-3)/2}M_1];$$

here M_1' is defined as in (2.6) for h rather than f_1.

tion ϕ of t

(3.1g) $\displaystyle\int u(x,t)\phi(t)\,dt = \iint h(s,\omega)\phi(x\omega - s)\,d\omega\,ds$

$$= [h(s,\omega),\,\phi(x\omega - s)],$$

where $h = (2\pi)^{(1-n)/2}(-\partial_s)^{(n-3)/2}k$. Thus, time like smoothing of even a distribution valued solution results in C^∞ data.

According to property (3.1d), $\mathcal{S}$ commutes with orthogonal transformations. This indicates that the spherical harmonic expansion may be useful in expressing some properties of $\mathcal{S}$. This is indeed the case:

Theorem 3.1. *The nullspace of $\mathcal{S}$ consists of all distributions k whose spherical harmonic expansion*

$$k = \sum k_{m,j}(s)\,Y_{m,j}(\omega)$$

has the following property: $k_{m,j}$ is a polynomial of degree less than $m + (n-3)/2$, where m is the order of the spherical harmonic $Y_{m,j}$.

Proof. Let $k_m(s)$ be a polynomial of degree less than $m + (n-3)/2$ and let Y_m be any spherical harmonic of order m. Since $k_m Y_m$ is C^∞, $u_m = W\mathcal{S}(k_m Y_m)$ is given by formula (2.21):

$$u_m(x,t) = \int h_m(x\omega - t)\,Y_m(\omega)\,d\omega$$

where

$$h_m = (2\pi)^{(1-n)/2}(-\partial_s)^{(n-3)/2}k_m .$$

Since $k_m(s)$ is a polynomial of degree less than $m + (n-3)/2$, $h_m(x\omega - t)$ is, for fixed x and t, a polynomial in ω of degree less than m; therefore, by the orthogonality properties of spherical harmonics, Y_m is orthogonal to $h_m(x\omega - t)$ and so u_m is zero for all x and t. But then so are the initial data $\mathcal{S}(k_m Y_m)$ of u_m .

Let k be any distribution whose spherical harmonic expansion $k = \sum k_{m,j}Y_{m,j}$ has the property stated in Theorem 3.1. Since $\mathcal{S}$ is continuous in the weak sequential topology,

$$\mathcal{S}k = \sum \mathcal{S}(k_{m,j}Y_{m,j}) .$$

We have shown above that each term on the right is zero; therefore so is $\mathcal{S}k$. This completes the proof of one part of the theorem.

To prove the converse, assume that $\mathfrak{g}k = 0$; then so is $u = W\mathfrak{g}k$. We now use the weak form (3.1g) of the representation formula for u which asserts that if $u = 0$ then for all C_0^∞ test functions $\phi(t)$

$$\iint h(s, \omega)\phi(x\omega - s)\, ds\, d\omega = 0$$

where h is the distribution $(2\pi)^{(1-n)/2}(-\partial_s)^{(n-3)/2}k$.

We apply $\partial_x{}^\alpha$ to both sides and set $x = 0$, obtaining

(3.2) $$\iint h(s, \omega)\omega^\alpha \partial_s{}^{|\alpha|}\phi(-s)\, ds\, d\omega = 0.$$

Using the abbreviation

(3.3) $$a_\alpha(s) = \int h(s, \omega)\omega^\alpha\, d\omega$$

and integrating by parts with respect to s we can rewrite (3.2) as

$$\int (\partial_s{}^{|\alpha|} a_\alpha(s))\phi(-s)\, ds = 0.$$

Since ϕ is an arbitrary C_0^∞ test function, it follows that $\partial_s{}^{|\alpha|} a_\alpha$ is zero, which implies that a_α is a polynomial of degree less than $|\alpha|$. Since the mth order spherical harmonic coefficients of h are linear combinations of expressions of the form (3.3) with $|\alpha| = m$, and since h is $\partial_s{}^{(n-3)/2}k$, it follows that k is of the form stated in the theorem.

The argument presented above can be easily modified to prove the following two corollaries:

Corollary 3.1. $\mathfrak{g}k$ *is zero for* $|x| < r$ *if and only if in the interval* $|s| < r$ *the spherical harmonic coefficient* k_m *of* k *is a polynomial of degree less than* $m + (n - 3)/2$.

Corollary 3.2. $\mathfrak{g}k$ *is outgoing (incoming) in the sense that* $W\mathfrak{g}k$ *is zero for* $|x| < t\,(|x| < -t)$ *if and only if* k_m *is a polynomial of degree less than* $m + (n - 3)/2$ *for* s *negative (positive)*.

As mentioned in Section 1, Corollary 3.1 furnishes us with a simple proof of the Holmgren uniqueness theorem for the wave equation:

Proof of Theorem 1.5. If $u(x, t)$ is a solution of the wave equation which vanishes inside the cylinder $|x| < R$, $|t| < T$, then according to the corollary the spherical harmonic coefficient $k_m(s - t)$ of the translation representer for $\{u(x, t), u_t(x, t)\}$ is a polynomial of degree less than $m + (n - 3)/2$ in $|s| < R$ for each t in $(-T, T)$ so that $k_m(s)$ is actually a polynomial of degree less than $m + (n - 3)/2$ in $(-T - R, T + R)$. The converse statement of the corollary now implies that $u(x, t)$ vanishes for $|x| < R + T - |t|$.

Likewise, Corollary 3.2 can be used to complete the proof of Theorem 2.3. We recall that Theorem 2.3 asserts that the translation representation $k = \mathcal{R}f$ of f in H_0 is zero on the negative (positive) s-axis if and only if f is outgoing (incoming). The only if part of this assertion was proved in Section 2.

Proof of the Converse Statement in Theorem 2.3. Suppose that f in H_0 is outgoing. Since f has finite energy, $k = \mathcal{R}f$ is square integrable; hence all spherical harmonic coefficients k_m of k are also square integrable. Since f is outgoing, according to Corollary 3.2 k_m is a polynomial for s negative and since k_m is square integrable, it follows that k_m is zero for s negative for all m; thus, k itself is zero for s negative.

The next result is in essence a dual to Theorem 3.1:

Theorem 3.2. *Let l be a distribution which is zero for $|s| > r$; then $\mathcal{G}l$ is zero for $|x| > r$ if and only if l satisfies the orthogonality conditions*

$$(3.4) \qquad\qquad [s^\beta Y_m(\omega), l] = 0$$

for all $\beta \leq m + (n - 3)/2$ and all spherical harmonics Y_m.

Proof. Assume that l satisfies the orthogonality conditions (3.4). Now the spherical harmonic coefficients of a test function ψ in C_0^∞ which is constant in s for say $|s| < r + \epsilon$ will also be constant in s in this interval and applying (3.4) with $\beta = 0$ to the expansion of ψ in spherical harmonics we see that $[l, \psi] = 0$. This permits us to define an indefinite integral p of the distribution l as

$$(3.5) \qquad\qquad [p, [\phi] = -[l, \psi]$$

where

$$\psi(s, w) = \int_{-\infty}^s \left\{ \phi(\sigma, \omega) - \left(\int_{-\infty}^\infty \phi(\tau, \omega) \, d\tau \right) \theta(\sigma, \omega) \right\} d\sigma$$

and $\theta(s, \omega)$ in C_0^∞ has its support in $|s| > r + \epsilon$ and is such that $\int \theta(s, \omega) \, ds \equiv 1$. In this case ψ is a test function with compact support so that p is well defined and, moreover,

$$[\partial_s p , \phi] = -[p , \partial_s \phi] = [l , \phi]$$

which shows that p is an indefinite integral of l. Finally, we note that the support of p is contained in $|s| \leq r$ since if ϕ has its support in $|s| > r + \epsilon$, then $\psi(s, \omega)$ will be constant in s for $|s| \leq r + \epsilon$ and hence as above $[p, \phi] = -[l, \psi] = 0$.

$$(3.6) \qquad\qquad\qquad (g, \mathcal{S}p)_E = [\Re g, p]$$

holds for every C_0^∞ function g. If the support of g lies outside the sphere $|x| \leq r + \epsilon$, then, according to Corollary 3.1 applied to $k = \Re g$, $\Re g$ can be expanded for $|s| \leq r + \epsilon$ into an infinite sum of terms of the form

$$\text{const } s^\alpha Y_m(\omega), \qquad \alpha < m + \frac{n-3}{2}.$$

Since g is C_0^∞, so is $k = \Re g$, and its spherical harmonics expansion converges in the C^∞ topology.

Now the support of p lies in $|s| \leq r$ and so we can write

$$[\Re g , p] = \sum \text{const } [s^\alpha Y_m(\omega) , p].$$

Applying formula (3.5) we can write each term on the right as a constant times $[(s^{\alpha+1} + \text{const}) Y_m(\omega), l]$; and since α is less than $m + (n-3)/2$ this will be zero by assumption (3.4). So it follows that $[\Re g, p]$ is zero and therefore according to the identity (3.6) so is $(g, \mathcal{S}p)_E$ for all C_0^∞ data g whose support lies outside the sphere $|x| \leq r$. Lemma 3.1 now asserts that

$$A_0 \mathcal{S}p = 0 \qquad \text{for} \quad |x| > r.$$

But by property (3.1b)

$$\mathcal{S}l = \mathcal{S}\partial_s p = -A_0 \mathcal{S}p ,$$

which proves one part of Theorem 3.2.

To prove the converse suppose that $\mathcal{S}l$ is zero for $|x| > a$, a some constant greater than r. For any given exponent β and spherical harmonic Y_j we construct an auxiliary function $\phi(s)$ with the following properties:

(3.7a) $\qquad \phi(s) = s^\beta \qquad$ for $\quad |s| < a + \epsilon$,

(3.7b) $\qquad \phi \quad$ is $\quad C_0^\infty$

(3.7c) $\qquad \phi(s) Y_j(\omega)$ satisfies the orthogonality conditions (3.4).

By the orthogonality of spherical harmonics, the function $\phi(s)Y_j(\omega)$ automatically satisfies all but a finite number of the conditions (3.4) for any choice of ϕ. The remaining finite number of conditions can easily be satisfied by suitably shaping ϕ outside the interval $|s| < a + \epsilon$.

Denote $\mathscr{I}\phi Y_j$ by g; we claim that if β satisfies

$$(3.8) \qquad\qquad \beta \le j + \frac{n-3}{2}$$

then

$$(3.9) \qquad\qquad A_0 g = 0 \quad \text{for} \quad |x| < a + \epsilon.$$

To see this we use property (3.1b) to write

$$(3.10) \qquad\qquad A_0 g = A_0 \mathscr{I}(\phi Y_j) = -\mathscr{I}\partial_s(\phi Y_j) \ .$$

By property (3.7a) $\partial_s\phi\, Y_j = \text{const } s^{\beta-1} Y_j$ for $|s| < a + \epsilon$, and hence (3.9) is an immediate consequence of Corollary 3.1.

It follows from (3.7c) and that part of Theorem 3.2 which we have already proved that $g = \mathscr{I}\phi Y_j$ has compact support. Also for ϕY_j in C_0^∞ it was proved in Section 2 that $\mathscr{I}$ and $\mathscr{R}$ are inverses of each other so that $\mathscr{R}g = \phi Y_j$.

We now apply (3.1f) to $k = l$:

$$(3.11) \qquad\qquad (g\,,\,\mathscr{I}l)_E = [\mathscr{R}g\,,\,l] = [\phi Y_j\,,\,l] \ .$$

By hypothesis $\mathscr{I}l$ vanishes for $|x| > a$ and according to (3.9) $A_0 g$ is zero for $|x| < a + \epsilon$. From this it follows by integration by parts (see Lemma 3.1) that $(g,\,\mathscr{I}l)_E = 0$ and hence by (3.11) that $[\phi Y_j,\,l] = 0$. Since the support of l is contained in the interval $(-r, r)$, and since by (3.7a) (and the relation $a > r$) $\phi = s^\beta$ there, it follows that

$$[\phi Y_j\,,\,l] = [s^\beta Y_j\,,\,l] = 0$$

for all β satisfying (3.8). In other words l satisfies the orthogonality conditions (3.4) This completes the proof of Theorem 3.2.

Notice that in the above proof of the necessity of the orthogonality conditions we only assumed that $\mathscr{I}l$ had compact support. Thus we have also proved

Corollary 3.3. *Let l be a distribution which is zero for $|s| > r$ and such that $\mathscr{g}l$ has compact support; then the support of $\mathscr{g}l$ is contained in the sphere $|x| \leq r$.*

In general $\mathscr{g}l$ need not have compact support with l. However, the behavior of $\mathscr{g}l$ for $|x| > r$ when l vanishes for $|s| > r$ is very restricted; in fact the spherical harmonic coefficients of such functions are analytic in $|x|$ for $|x| > r$.

Theorem 3.3. *Let l be a distribution which is zero for $|s| > r_0$ and set $u = W\mathscr{g}l$. If $u_m(t)$ denotes the mth spherical harmonic coefficient of $u(t)$:*

$$u_m(t) = (u(t), Y_m(\theta)),$$

then u_m is an analytic function of $r = |x|$ and t for $|r| > |t| + r_0$, and for fixed $r > r_0$ it can be extended to be analytic in t in the cut plane excluding the set of real t: $t > r - r_0$ and $t < -r + r_0$.

Proof. We begin with the relation (3.1g) according to which

$$\int u(t)\phi(t)\,dt = \iint h(s,\omega)\phi(x\omega - s)\,d\omega\,ds$$

for all ϕ in $C_0^\infty(R)$; here $h(s,\omega)$, which is given by (2.9), also has its support in $|s| \leq r_0$. The spherical harmonic coefficient of $u(t)$ therefore satisfies the relation

$$(3.12) \qquad \int u_m(t)\phi(t)\,dt = \iiint h(s,\omega)\phi(r\theta\omega - s)Y_m(\theta)\,d\omega\,d\theta\,ds.$$

We can partially evaluate (3.12) by rotating the θ coordinate system so that the first coordinate is in the ω-direction. It is easy to show† that there

---

† For each ω_0 construct an orthonormal basis $\{\omega_0, e_2, \ldots, e_n\}$. Then the set of vector-valued functions

$$f_1(\omega) = \omega,$$

$$f_j(\omega) = e_j, \qquad (j = 2, 3, \ldots, n,)$$

are orthonormal at $\omega = \omega_0$ and hence linearly independent for ω sufficiently close to ω_0. In this neighborhood of ω_0 we can use the f_j's to construct an orthonormal basis of the type (3.13) by the Gram–Schmidt procedure. The resulting basis is a continuous function of ω in this neighborhood of ω_0. Using the set of all such neighborhoods to provide a covering for S_{n-1} and extracting a finite subcovering, we can define a family of bases on all of S_{n-1} which is piecewise continuous and *a fortiori* Borel measurable.

exists a Borel measurable family of orthonormal bases for R_n defined on S_{n-1} of the form

(3.13) $\{e_1(\omega) \equiv \omega , e_2(\omega) , \cdots , e_n(\omega) \}.$

Employing this basis we can write

$$\theta = \sum_{j=1}^{n} \xi_j e_j(\omega)$$

so that the original θ coordinates (relative to say $\{e_1{}^0, e_2{}^0, \cdots, e_n{}^0\}$) become

$$\theta_k = \theta e_k{}^0 = \sum_{j=1}^{n} \xi_j (e_j(\omega) e_k{}^0).$$

We note that the $e_j(\omega) e_k{}^0$ are bounded measurable functions of ω. Since $Y_m(\theta)$ is a homogeneous polynomial in θ of degree m, it can now be written as
$$Y_m(\theta) = \sum_{|\alpha|=m} a_\alpha(\omega) \xi^\alpha ,$$

where again the $a_\alpha(\omega)$ are bounded measurable functions. The relation (3.12) becomes

(3.14) $\displaystyle \int u_m(t) \phi(t) \, dt = \iiint \sum_\alpha a_\alpha(\omega) h(s, \omega) \xi^\alpha \phi(r\xi_1 - s) \, d\omega \, d\xi \, ds.$

Denoting the last $(n - 1)$ ξ coordinates by ξ', we can perform the ξ' integration over the $(n - 2)$-sphere of radius $(1 - \xi_1{}^2)^{1/2}$. Recalling that $d\xi = (1 - \xi_1{}^2)^{-1/2} \, d\xi_1 \, d\xi'$ and setting $\alpha = \{\alpha_1, \alpha'\}$, the integration with respect to ξ' of a generic term in (3.14) is

$$\int_{|\xi|=1} g(\xi_1) \xi^\alpha \, d\xi = \int_{-1}^{1} g(\xi_1) \xi_1{}^{\alpha_1} (1 - \xi_1{}^2)^{-1/2} \left\{ \int_{|\xi'|=(1-\xi_1{}^2)^{1/2}} (\xi')^{\alpha'} \, d\xi' \right\} d\xi_1$$

$$= \left\{ \int_{|\xi'|=1} (\xi')^{\alpha'} \, d\xi' \right\} \int_{-1}^{1} g(\xi_1) \xi_1{}^{\alpha_1} (1 - \xi_1{}^2)^{(n-3+|\alpha'|)/2} \, d\xi_1 .$$

Consequently, we can express (3.14) in the form

$$\int u_m(t) \phi(t) \, dt$$
$$= \iiint \sum_\alpha b_\alpha(\omega) h(s, \omega) \xi_1{}^{\alpha_1} (1 - \xi_1{}^2)^{(n-3+|\alpha'|)/2} \phi(r\xi_1 - s) \, d\omega \, d\xi_1 \, ds$$

where the $b_\alpha(\omega)$ are also bounded measurable functions. A simple change of variable gives

$$\int u_m(t)\phi(t)\,dt = \int \left\{\frac{1}{r} \iint \sum_\alpha b_\alpha(\omega)h(s,\omega)\left(\frac{t+s}{r}\right)^{\alpha_1}\right.$$
$$\left.\times \left[1-\left(\frac{t+s}{r}\right)^2\right]^{(n-3+|\alpha'|)/2} d\omega\,ds\right\}\phi(t)\,dt.$$

Now $h(s,\omega)$ has its support in $|s| \leq r_0$ and in this range the factor of $h(s,\omega)$ in the above expression is smooth, in fact analytic for $|r| > |t| + r_0$. Hence, considered as a distribution on ϕ, $u_m(t)$ can be represented by a function of r and t:

$$(3.15) \quad u_m(r,t) = \sum_\alpha \frac{1}{r}\iint b_\alpha(\omega)h(s,\omega)\left(\frac{t+s}{r}\right)^{\alpha_1}$$
$$\times \left[1-\left(\frac{t+s}{r}\right)^2\right]^{(n-3+|\alpha'|)/2} d\omega\,ds$$

and it is obvious that this expression has all of the required analytic properties.

We note that Corollary 3.3 can also be proved from Theorem 3.3 since it follows from the analyticity of the spherical harmonic coefficients that they vanish identically for $r > r_0$ if $\mathcal{I}$ has compact support.

4. Translation Representation for Outgoing and Incoming Data with Infinite Energy

In Section 2 we defined the translation representation $\mathcal{R}f$ of any C_0^∞ data f, and then extended the operator $\mathcal{R}$ by continuity to all data f with finite energy. In this section we shall give further extensions of $\mathcal{R}$.

First of all we observe that $\mathcal{R}$ can be extended by continuity to all distributions g with compact support. It suffices to approximate g by a sequence $\{g_n\}$ of C_0^∞ data with support in a fixed compact set; then making use of the explicit formula (2.5) for $\mathcal{R}g_n$ and the relation

$$(g_n,\mathcal{I}k)_E = [\mathcal{R}g_n, k]$$

which holds for all C_0^∞ functions k, we see that $\mathcal{R}g_n$ tends weakly to a distribution which is defined to be $\Re g$. Comparing this with the definition of $\mathscr{I}$ we see at once that $\mathscr{I}\Re g = g$.

Remark 4.1. If g is a distribution with compact support then $\Re g$ is a C^∞-valued distribution in s; that is for any ϕ in $C_0^\infty(R_1)$

$$\int \Re g\phi(s) \ ds$$

is a C^∞ function of ω. This is easy to verify for C_0^∞ data g because in this case formula (2.5) gives

$$\int \Re g\phi(s) \ ds = (-1)^{(n-1)/2} \int [M_1(s, \omega)\phi^{(n+1)/2}(s)$$
$$+ M_2(s, \omega)\phi^{(n-1)/2}(s)] \ ds ,$$

and applying (2.6) we get

$$\int \Re g\phi(s) \ ds = \text{const} \int [g_1(x)\phi^{(n+1)/2}(x\omega) + g_2(x)\phi^{(n-1)/2}(x\omega)] \ dx .$$

The latter expression holds by continuity for distribution data with compact support and since the right side is obviously C^∞ in ω this proves the assertion.

An equivalent definition can be based directly on formula (3.1f):

(4.1) $$(g , \mathscr{I}k)_E = [\Re g , k] .$$

Since g has compact support, the left side is meaningful for any C_0^∞ function k and it is easy to show that it depends continuously on k. Thus (4.1) defines $\Re g$ as a distribution.

It is easy to verify that the two definitions are equivalent. It follows from either definition that $\Re$ has the following properties:

(4.2a) If $g = 0$ for $|x| > r$, $\Re g = 0$ for $|s| > r$.

(4.2b) $\Re A_0 = -\partial_s \Re$.

(4.2c) $\mathscr{I}\Re = I$.

As an example, we compute $\Re\{0, \delta\}$: according to formula (2.7), the value of the second component of gk at $x = 0$ is given by

$$[gk]_2(0) = \frac{1}{(2\pi)^{(n-1)/2}} \int (-\partial_s)^{(n-1)/2} k(0, \omega) \, d\omega.$$

Comparing this with (4.1) we conclude that

$$(4.3) \qquad \mathfrak{R}\{0, \delta\} = \frac{1}{2(2\pi)^{(n-1)/2}} \partial_s^{(n-1)/2} \delta(s)$$

where $\delta(s)$ denotes the one-dimensional delta function.

One way of further extending the operator $\mathfrak{R}$ is to consider all distributions g of the form gl and then define $\mathfrak{R}g$ as l. The difficulty with this approach is that it may not be easy to decide if a given g belongs to the range of g; also, in this way $\mathfrak{R}g$ would be defined only modulo the nullspace of g. In this section we shall employ instead a special method to define $\mathfrak{R}$ for a restricted class of distributions which are particularly relevant to scattering theory. This class is made up of *eventually outgoing* data, defined as follows.

Definition 4.1. f is called eventually outgoing if there is a constant r such that $U_0(r)f$ is outgoing, i.e., Wf is zero for $|x| < t - r$.

Initially incoming is defined similarly.

It follows from formula (3.1g) that if k is zero for $s < -r$ then gk is eventually outgoing; a partial converse is given in Corollary 3.2.

Theorem 4.1. *Let g be any data with compact support and μ any complex number; then the equation*

$$(4.4) \qquad (A_0 - \mu)f = g$$

has a unique solution f which is eventually outgoing (initially incoming).

Proof. Applying $\mathfrak{R}$ formally to (4.4) and using (4.2b) yields the following ordinary differential equation for $k = \mathfrak{R}f$:

$$(4.5) \qquad -(\partial_s + \mu)k = \mathfrak{R}g.$$

For smooth data g with compact support, $\mathfrak{R}g$ is a function with compact support and a solution of (4.5) which vanishes near $-\infty$ can be written as

$$(4.6)' \qquad k(s) = -\int_{-\infty}^{s} e^{\mu(\sigma-s)} \mathfrak{R}g(\sigma) \, d\sigma.$$

However, for distribution data with support in $\{|x| \leq r\}$ we have to express the indefinite integral indirectly as in (3.5). In this case we obtain

$$(4.6) \quad [k, \phi] = \left[\Re g, \, \exp\,(\bar{\mu}s) \int_{-\infty}^{s} \left\{ \exp\,(-\bar{\mu}\sigma)\phi(\sigma, \omega) \right. \right.$$

$$\left. \left. - \left(\int_{-\infty}^{\infty} \exp\,(-\bar{\mu}\tau)\phi(\tau, \omega) \, d\tau \right) \theta(\sigma, \omega) \right\} d\sigma \right],$$

where θ is chosen from C_0^{∞} so as to have its support in $(-\infty, -r)$ and so that $\int \theta(s, \omega) \, ds \equiv 1$. We claim that $f = \vartheta k$ solves our problem. For by assumption g is zero outside the ball $|x| \leq r$. Thus it follows from (4.2a) that $\Re g$ is zero for $|s| > r$ and so it follows from (4.6) that k is zero for $s < -r$. This shows that $f = \vartheta k$ is eventually outgoing.

Using (3.1b), (4.5), and (4.2c) we have

$$(A_0 - \mu)f = (A_0 - \mu)\vartheta k = -\vartheta(\partial_s + \mu)k = \vartheta \Re g = g$$

which proves that f is a solution of (4.4).

To prove the uniqueness of the solution suppose that there are two; then their difference g is also outgoing and satisfies the homogeneous equation

$$(4.7) \qquad\qquad A_0 g = \mu g \, .$$

Consider now $U_0(t)g$; differentiating it with respect to t and using (4.7) we get

$$\frac{d}{dt} U_0(t)g = U_0(t)A_0 g = \mu U_0(t)g \, .$$

Integrating this ordinary differential equation we get

$$(4.8) \qquad\qquad U_0(t)g = e^{\mu t}g \, .$$

Since g is eventually outgoing, $U_0(t)g$ is zero for $|x| < t - r$; so it follows from (4.8) that $g(x) = 0$ for $|x| < t - r$. Since t is arbitrary, it follows that $g \equiv 0$, this proves that solutions are unique and completes the proof of Theorem 4.1.

In the course of proving Theorem 4.1 we have shown that the solution f has a translation representation k. We summarize the properties of k in

Corollary 4.1. *Let f be eventually outgoing (initially incoming) and suppose that for some complex number* μ

$$(A_0 - \mu)f$$

is zero for $|x| > r$. *Then f has a translation representer k called outgoing (incoming) with the following properties:*

$$(4.9)_o \qquad k(s, \omega) = \begin{cases} 0 & \text{for} \quad s < -r \\ e^{-\mu s} n(\omega) & \text{for} \quad s > r \end{cases} .$$

In the incoming case:

$$(4.9)_i \qquad k(s, \omega) = \begin{cases} e^{-\mu s} n(\omega) & \text{for} \quad s < -r \\ 0 & \text{for} \quad s > r) \end{cases} ;$$

here $n(\omega)$ *lies in* $C^\infty(S_{n-1})$.

Proof. Since $\Re g$ vanishes for $|s| > r$ by (4.2a), formula (4.6) can be written for ϕ with support in (r, ∞) as

$$[k, \phi] = - \left[\Re g, \exp(\bar{\mu}s) \left(\int_{-\infty}^{\infty} \exp(-\bar{\mu}\tau) \phi(\tau, \omega) \, d\tau \right) \right]$$

$$= - \iint \Re g e^{\mu s} \left(\int_{-\infty}^{\infty} e^{-\mu\tau} \overline{\phi(\tau, \omega)} \, d\tau \right) ds \, d\omega = [e^{-\mu\tau} n(\omega), \phi]$$

where we have set

$$n(\omega) = - \int \Re g e^{\mu s} \, ds.$$

According to Remark 4.1 the function $n(\omega)$ is C^∞ in ω. This proves $(4.9)_o$ and $(4.9)_i$ follows similarly.

Remark 4.2. If instead of assuming that f is eventually outgoing or initially incoming, we had merely assumed that f has a translation representer k and that $(A_0 - \mu)f = g$ vanishes for $|x| > r$, then we could argue as follows: By (4.2a) $\Re g$ vanishes for $|s| > r$ and hence for any test function ϕ with support in, say $s > r$, we have

$$0 = [(\partial_s + \mu)k, \exp(\bar{\mu}s)\phi] = -[k, \exp(\bar{\mu}s)\partial_s\phi] = [\partial_s e^{\mu s}k, \phi].$$

Consequently, for $s > r$, $e^{\mu s}k$ is independent of s and hence is a distribution in ω alone. A similar argument holds for $s < -r$ so that

$$(4.9) \qquad k(s, \omega) = \begin{cases} e^{-\mu s} n_1(\omega) & \text{for} \quad s < -r, \\ e^{-\mu s} n_2(\omega) & \text{for} \quad s > r \;. \end{cases}$$

Corollary 4.2. *Let f and g be eventually outgoing and initially incoming, respectively, such that $(A_0 - \mu)f$ and $(A_0 + \bar{\mu})g$ have compact support. Suppose that f is C^∞, g a distribution. Then*

$$(4.10) \qquad -((A_0 - \mu)f, g)_E = (f, (A_0 + \bar{\mu})g)_E\;.$$

Proof. Denote by k and l the outgoing and incoming representations of f and g, respectively. In this case $(4.6)'$ makes sense and shows that k is smooth. By the customary properties of $\mathscr{I}$ and $\mathscr{R}$ we can write the left and right side of (4.10) as

$$[(\partial_s + \mu)k\,,\, l] \qquad \text{and} \qquad [k\,,\, (-\partial_s + \bar{\mu})l]\;.$$

But these two are equal since they can be transformed into each other by an integration by parts; there are no boundary terms since k is zero for large negative s, l for large positive s.

In Chapter V we shall consider an analogue of Theorem 4.1, where the functions f and g are defined only outside a compact domain on whose boundary f is required to be zero. We shall prove that this problem has a unique solution except for a denumerable set of μ. The exceptional values of μ turn out to be the poles of the scattering matrix.

The following is a useful consequence of Corollary 4.1:

Theorem 4.2. *Suppose that f is both eventually outgoing and initially incoming, and that for some complex number μ not equal to zero $(A_0 - \mu)f$ is zero for $|x| > r$. Then f itself is zero for $|x| > r$.*

Proof. By Corollary 4.1, f has incoming and outgoing translation representations k_i and k_o, satisfying $(4.9)_i$ and $(4.9)_o$, respectively. We claim that the two are equal; for $k_o - k_i$ represents zero; therefore, by Theorem 3.1 its spherical harmonic coefficients are polynomials in s. On the other hand it follows from $(4.9)_o$ and $(4.9)_i$ that $k_o - k_i$ is an exponential function of s for both $s > r$ and $s < -r$; but then so are the coefficients of $k_o - k_i$. However, the only function which is both a polynomial and an exponential is a constant, and since $\mu \neq 0$ that constant has to be zero. This proves that all coefficients of $k_o - k_i$ are zero, and therefore $k_o - k_i$ itself is zero.

Set $k = k_o = k_i$. It follows from $(4.9)_o$ and $(4.9)_i$ that k is zero for $|s| > r$. By construction $(-\partial_s - \mu)k$ is a translation representation for $(A_0 - \mu)f$ which is by assumption zero for $|x| > r$. Since $\mathcal{G}(\partial_s + \mu)k$ has compact support, it follows from Theorem 3.2 that $(\partial_s + \mu)k$ satisfies the orthogonality conditions (3.4):

$$[s^\beta Y_m , \partial_s k + \mu k] = 0 \qquad (\beta \leq m + (n - 3)/2) .$$

Integrating this by parts gives

(4.12) $$-\beta[s^{\beta-1}Y_m , k] + \mu[s^\beta Y_m , k] = 0 .$$

We claim that

(4.13) $$[s^\beta Y_m , k] = 0 \qquad (\beta \leq m + (n - 3)/2) .$$

For $\beta = 0$ this follows directly from (4.12) and for higher values of β it follows by induction also from (4.12), all under the assumption that $\mu \neq 0$. The relation (4.13) shows that k satisfies the orthogonality conditions (3.4); so by the first part of Theorem 3.2 $\mathcal{G}k$ is zero for $|x| > r$, as asserted in Theorem 4.2.

Corollary 4.3. *Let f denote data of finite energy and suppose for some real σ not equal to zero that $(A_0 - i\sigma)f = 0$ for $|x| > r$. Then f vanishes for $|x| > r$.*

Proof. Since f has finite energy it has a square integrable translation representation k. According to the remark following Corollary 4.1, $k(s, \omega)$ is of the form $\exp(-i\sigma s)n_1(\omega)$ for $s < -r$ and of the form $\exp(-i\sigma s)n_2(\omega)$ for $s > r$. However, since k is square integrable both n_1 and n_2 must be zero. Thus, f is both eventually outgoing and initially incoming and since $\sigma \neq 0$, Theorem 4.2 implies that f itself is zero for $|x| > r$.

An example. Take $g = \{0, \delta\}$ in Eq. (4.4); to find f we have to solve the ordinary differential equation (4.5). According to (4.3)

$$\mathcal{R}g = \frac{1}{2(2\pi)^{(n-1)/2}} (\partial_s)^{(n-1)/2} \delta(s) ;$$

the solution of (4.5) can then be explicitly computed from (4.6) as:

(4.14) $$k_\mu(s) = -\frac{1}{2(2\pi)^{(n-1)/2}} \left\{ \sum_{\alpha=1}^{(n-1)/2} (-\mu)^{(n-1)/2-\alpha} \partial_s^{\alpha-1} \delta(s) \right.$$
$$\left. + (-\mu)^{(n-1)/2} \eta(s)e^{-\mu s} \right\} ,$$

where η is the unit function

$$\eta(s) = \begin{cases} 0 & \text{for } s < 0 \\ 1 & \text{for } s \geq 0 \end{cases} .$$

To find f_μ itself we have to evaluate $\mathscr{g}k_\mu$; since k_μ is independent of ω, $\mathscr{g}k_\mu$ is a function of $|x|$ alone and all but one of the integrations in formula (2.8) can be carried out explicitly:

$$(4.15) \quad (\mathscr{g}k)_1 = (-1)^{(n-3)/2} \frac{1}{(2\pi)^{(n-1)/2}} \int k^{(n-3)/2}(x\omega) \, d\omega$$

$$= (-1)^{(n-3)/2} \frac{\omega_{n-1}}{(2\pi)^{(n-1)/2}} \int_0^\pi k^{(n-3)/2}(r \cos\theta)(\sin\theta)^{n-2} \, d\theta$$

$$= (-1)^{(n-3)/2} \frac{\omega_{n-1}}{(2\pi)^{(n-1)/2}} \int_{-1}^1 k^{(n-3)/2}(r\rho)(1 - \rho^2)^{(n-3)/2} \, d\rho$$

$$= (-1)^{(n-3)/2} \frac{\omega_{n-1}}{(2\pi)^{(n-1)/2}} \frac{1}{r^{n-2}} \int_{-r}^r k^{(n-3)/2}(\tau)(r^2 - \tau^2)^{(n-3)/2} \, d\tau .$$

Setting $k = k_\mu$ as given by formula (4.14) we obtain, after a brief calculation†

$$(4.16) \quad \gamma_\mu = [\mathscr{g}k_\mu]_1 = \frac{(-1)^{(n-1)/2}}{2(2\pi)^{(n-1)/2}} \left(\frac{1}{r} \partial_r\right)^{(n-3)/2} \left(\frac{e^{-\mu r}}{r}\right) .$$

† Denoting $k_\mu(s)$ for n dimensions by $k(s, n)$, one readily verifies from (4.14) that

$$k^{(1)}(s, n) = 2\pi k(s, n+2).$$

Hence

$$\frac{1}{r} \frac{\partial}{\partial r} \int_{-1}^1 k^{(n-3/2)}(r\rho, n)(1 - \rho^2)^{(n-3)/2} \, d\rho = \frac{1}{r} \int_{-1}^1 k^{(n-1)/2}(r\rho, n)\rho(1 - \rho^2)^{(n-3)/2} \, d\rho$$

$$= \frac{-2\pi}{(n-1)r} \int_{-1}^1 k^{(n-3)/2}(r\rho, n+2)d(1 - \rho^2)^{(n-1)/2}$$

$$= \frac{2\pi}{n-1} \int_{-1}^1 k^{(n-1)/2}(r\rho, n+2)(1 - \rho^2)^{(n-1)/2} \, d\rho.$$

The result now follows by induction.

Next we consider arbitrary data g with compact support whose first component is zero:

$$g = \{0, w\} \ .$$

Equation $(A_0 - \mu)f = g$ can then be conveniently expressed in terms of the components of f as follows:

$$f_2 = \mu f_1, \qquad \Delta f_1 - \mu f_2 = w \ .$$

Denoting the first component f_1 by v, we can write f in the form

(4.17) $$f = \{v \, , \mu v\} \ ,$$

and v satisfies the *reduced wave equation*

(4.18) $$\Delta v - \mu^2 v = w \ .$$

Definition 4.2.   A solution v of the inhomogeneous reduced wave equation (4.18), where w has compact support, is called μ-*outgoing* (μ-*incoming*) if the data (4.17) are eventually outgoing (initially incoming).

It follows from Theorem 4.1 that given any distribution w with compact support, Eq. (4.18) has exactly one μ-outgoing and one μ-incoming solution. For $w = \delta$ the μ-outgoing solution γ_μ is, according to our calculations, given by formula (4.16); the μ-incoming solution is obtained by replacing μ by $-\mu$ since the incoming solution of (4.4) with $g = \{0, \delta\}$ is $f = \{\gamma_{-\mu}, \mu\gamma_{-\mu}\}$. These functions are called the μ-*outgoing and μ-incoming Green's functions* of the reduced wave equation.

The μ-outgoing solution of (4.18) for arbitrary w is the convolution of w and γ_μ :

(4.19) $$v(x) \ = w * \gamma_\mu = \iint w(y)\gamma_\mu(x - y) \, dy \ .$$

Choose any domain G with a smooth boundary which contains the support of w; then the integration with respect to y on the right in (4.19) may be restricted to G. Suppose now that x lies outside of G; then if we express w in the integrand in (4.19) as $(\Delta - \mu^2)v$, integrate by parts and make use of the differential equation satisfied by γ_μ we obtain the following formula for $v(x)$:

$$(4.20) \qquad v(x) = \int_{\partial G} \left[\frac{\partial v(y)}{\partial n} \gamma_\mu(x-y) - v(y) \frac{\partial \gamma_\mu(x-y)}{\partial n} \right] dS,$$

where n denotes the outward normal to ∂G, the boundary of G.

Theorem 4.3. *Let v be a solution of the homogeneous reduced wave equation outside some bounded smooth domain G. Then v is μ-outgoing if and only if the representation formula (4.20) holds outside of G, where γ_μ is given by formula (4.16).*

According to Huygens' principle, the μ-outgoing property of initial data is not effected by a change on any compact set. Hence it suffices to define v inside G in any smooth manner, in which case the derivation of (4.20) for μ-outgoing v constitutes the proof of the only if part. That any v represented by (4.20) is μ-outgoing is obvious since then v is a superposition of μ-outgoing solutions γ_μ and $\partial \gamma_\mu/\partial n$.

The representation (4.20) for v is useful in determining the asymptotic behavior of $v(x)$ for $|x|$ large. The leading term of γ_μ is

$$(4.21) \qquad \gamma_\mu(x) = \frac{(-1)^n \mu^{(n-3)/2}}{2(2\pi)^{(n-1)/2}} \frac{e^{-\mu r}}{r^{(n-1)/2}} + O\left(\frac{e^{-\mu r}}{r^{(n+1)/2}} \right) \qquad (r = |x|).$$

Denoting x/r by θ, an easy calculation gives

$$\gamma_\mu(x-y) = \frac{(-1)^n \mu^{(n-3)/2}}{2(2\pi)^{(n-1)/2}} \frac{e^{-\mu r}}{r^{(n-1)/2}} \exp(\mu\theta y) + O\left(\frac{e^{-\mu r}}{r^{(n+1)/2}} \right),$$

substituting these asymptotic formula for γ_μ into (4.20) we obtain an asymptotic formula for v:

$$(4.21)' \qquad\qquad v(x) = \frac{e^{-\mu r}}{r^{(n-1)/2}} m(\theta) + O\left(\frac{e^{-\mu r}}{r^{(n+1)/2}} \right),$$

m some function of θ; the integral representation for m shows that m is an analytic function of θ. Furthermore, formal differentiation of $(4.21)'$ gives correctly the asymptotic behavior of the derivatives of v.

For $\mathrm{Re}\,\mu \geq 0$ the asymptotic behavior $(4.21)'$ characterizes μ-outgoing solutions. We do not even need to use the pointwise behavior of v as given by (4.21)', only upper bounds for the following integrals over large spheres:

$$(4.22)_1 \qquad \int_{|x|=r} |v| \, dS = o(r^{(n+1)/2})$$

and

$$(4.22)_2 \qquad \int_{|x|=r} \left| \mu v + \frac{\partial v}{\partial r} \right| dS = o(r^{(n-1)/2}) \ ;$$

here dS is integration with respect to surface area. Clearly, if v has the asymptotic behavior (4.21)' and if $\partial v / \partial r$ has the asymptotic behavior obtained by differentiating (4.21) formally, then $(4.22)_1$ and $(4.22)_2$ are satisfied for Re $\mu \geq 0$.

Conversely:

Theorem 4.4. *Let v be a solution of the reduced wave equation outside of some bounded domain G, which satisfied (4.22) for large r, and suppose that Re $\mu \geq 0$; then v is μ-outgoing.*

Proof. Let x be any point outside of G; choose r so large that the sphere S_r around x contains G in its interior. Denote by H_r the region outside G and inside S_r. Then

$$v(x) = \iint_{H_r} v(y) \, \delta(x-y) \, dy = \iint_{H_r} v(y) \, (\Delta^2 - \mu^2) \gamma_\mu (x-y) \, dy \ .$$

Integrating by parts and using the fact that $(\Delta - \mu^2)v = 0$ outside G we obtain

$$(4.23) \quad v(x) = \int_{\partial G} \left[\frac{\partial v}{\partial n} \gamma_\mu - v \frac{\partial \gamma_\mu}{\partial n} \right] dS - \int_{S_r} \left[\frac{\partial v}{\partial n} \gamma_\mu - v \frac{\partial \gamma_\mu}{\partial n} \right] dS \ .$$

The explicit formula (4.16) shows that

$$\frac{\partial}{\partial n} \gamma_\mu = -\mu \gamma_\mu + O \left(\frac{1}{r^{(n+1)/2}} \right) .$$

Substituting this into (4.23), using the estimates (4.22) for v and the asymptotic behavior of γ_μ as given by (4.21), we see that the second integral on the right in (4.23) tends to zero as r tends to infinity. This shows that $v(x)$ has the representation (4.20), which according to the trivial part of Theorem 4.3 implies that v is μ-outgoing.

We have shown that every μ-outgoing solution v of the reduced wave equation outside of some bounded set has the asymptotic behavior described by (4.21)':

$$v(x) = \frac{e^{-\mu r}}{r^{(n-1)/2}} m(\theta)[1 + o(1)],$$

where $m(\theta)$ is an analytic function of $\theta = x/|x|$. According to Corollary 4.1 the translation representation k of $f = \{v,\, \mu v\}$ is, for s large enough, equal to

$$k(s,\, \omega) = e^{-\mu s} n(\omega).$$

We now show that $n(\omega)$ is essentially equal to $m(\omega)$.

Theorem 4.5. *For every μ-outgoing solution of the reduced wave equation*

$$(4.24) \qquad m(\omega) = \frac{(-1)^{(n-1)/2}}{\mu} n(\omega).$$

Proof. We shall verify this for the outgoing Green's function

$$v(x) = \gamma_\mu(x - y) = \frac{(-1)^n \mu^{(n-3)/2}}{2(2\pi)^{(n-1)/2}} \frac{e^{-\mu|x-y|}}{|x - y|^{(n-1)/2}} + \cdots$$

and its y-derivative. According to (4.20) every outgoing solution can be represented as a superposition of such functions, so that this will suffice to prove the theorem.

Using the asymptotic relation

$$|x - y| = (x^2 - 2xy + y^2)^{1/2} = |x|\left(1 - 2\frac{xy}{|x|^2} + \frac{y^2}{|x|^2}\right)^{1/2}$$

$$= |x| - \frac{x}{|x|}y + O\left(\frac{1}{|x|}\right),$$

we obtain

$$(4.25) \quad \gamma_\mu(x - y) = \frac{(-1)^n \mu^{(n-3)/2}}{2(2\pi)^{(n-1)/2}} \frac{e^{-\mu|x|}}{|x|^{(n-1)/2}} \exp\left(\mu \frac{x}{|x|} y\right) + \cdots.$$

On the other hand the translation representation of $f_\mu = \{\gamma_\mu(x),\, \mu\gamma_\mu(x)\}$

is, according to (4.14), given by

$$k_\mu(s, \omega) = -\frac{1}{2(2\pi)^{(n-1)/2}}(-\mu)^{(n-1)/2}e^{-\mu s} \quad \text{for} \quad s > 0.$$

The translation representation of $f_\mu(x - y)$ is simply $k_\mu(s - y\omega, \omega)$; so that

$$(4.26) \quad k_\mu(s - y\omega, \omega) = -\frac{1}{2(2\pi)^{(n-1)/2}}(-\mu)^{(n-1)/2}e^{-\mu s}e^{\mu y\omega} \quad \text{for} \quad s > |y|.$$

The relations (4.25) and (4.26) show that the functions $m(\omega)$ and $n(\omega)$ associated with $v(x) = \gamma_\mu(x - y)$ are

$$m(\omega) = \frac{(-1)^n \mu^{(n-3)/2}}{2(2\pi)^{(n-1)/2}}e^{\mu y\omega}, \qquad n(\omega) = \frac{(-1)^{(n+1)/2}\mu^{(n-1)/2}}{2(2\pi)^{(n-1)/2}}e^{\mu y\omega}$$

which verifies (4.24) in this case. Finally, it is readily verified that the effect of a y-differentiation in the n-direction is to introduce a factor of $\mu\omega n$ in both of the above expressions.

A similar calculation for μ-incoming solutions of the reduced wave equation gives:

$$(4.27) \qquad\qquad m(\omega) = \frac{1}{\mu}n(-\omega).$$

5. Notes and Remarks

The material in this chapter is basically classical. However, our point of view is somewhat novel and we have tailored the theory to fit the specific needs of our approach to the scattering problem for the wave equation. The translation representation is of course the Radon transform in a form which is natural for this situation. Very possibly our way of looking at the Radon transform may be useful in generalizing it to other situations.

For further treatments of the Radon transform and its generalizations see Helgason [1], John [1], and Gelfand et al. [1].

Theorem 2.4 goes back to Friedlander [1]; however, our proof is new.

There are many questions left unanswered by our treatment of the Radon transform. For instance we have not been able to characterize the class of data having translation representers. The following argument shows that there exist eventually outgoing data not having translation representers which vanish on a negative half-line. Suppose on the contrary that each data f such that $u = Wf$ is zero for $|x| < t$ has a translation representer $h(s, \omega)$ which vanishes for s sufficiently negative, and hence by Corollary 3.1 for $s < 0$. Let u be any solution of the wave equation which vanishes in a half-space:

$$(5.1) \qquad\qquad u(x, t) = 0 \qquad \text{for} \quad x_1 < t \ ;$$

then for all y such that $y_1 > 0$

$$u_y(x, t) \equiv u(x - y, t) \text{ is 0 for} \quad |x| < t .$$

An obvious translation representer for the initial data of u_y is

$$h_y(s, \omega) = h(s - y\omega, \omega)$$

and since h_y vanishes for $s < -|y|$, Corollary 3.1 requires that h_y be zero for all $s < 0$. Setting $t = s - y\omega$ we see that

$$h(t, \omega) = 0$$

for all $\{t, \omega\}$ such that $t + y\omega < 0$ for some y with $y_1 > 0$. Clearly, the set of such $\{t, \omega\}$ include all pairs except $\{t \geq 0, \omega = \omega_1 \equiv (1, 0, \cdots, 0)\}$. It follows† that h is a locally finite sum of the type

$$\sum h_\alpha(s) \partial_\omega{}^\alpha \delta(\omega - \omega_1),$$

where the $h_\alpha(s)$ vanish for $s < 0$. Consequently, u is a locally finite sum:

$$\sum x^\alpha f_\alpha(x_1 - t),$$

where the $f_\alpha(t)$ are suitably related and vanish for $t < 0$. On the other hand it is easy to construct other solutions to the wave equation which satisfy (5.1); the following was suggested by Reubin Hersch:

$$u(x, t) = \phi(x_2, \cdots, x_n) f(x_1 - t)$$

where ϕ is harmonic and f is arbitrary but with support on the positive axis.

---

† We may suppose without loss of generality that $h(s, \omega)$ is smooth in s weakly since we can attain this in any case by smoothing in the time direction.

CHAPTER V

The Solution of the Wave Equation in an Exterior Domain

All of the preceding material has been developed for the purpose of studying the wave equation in an exterior domain; this we now do. Essentially what remains to be done is to fit this problem into the abstract framework and to verify the hypotheses of some of the previous theorems.

In Section 1 we consider weak solutions of the wave equation defined for all time and for all x in the exterior of some obstacle on whose boundary the solutions are required to be zero. The initial data can be prescribed arbitrarily and if they have finite energy the solution will have the same energy at all other times. The operator mapping initial data into data at time t is denoted by $U(t)$; these operators form a one-parameter group which is unitary in the energy norm. Our proof of these familiar facts of wave propagation consists in showing that the formal infinitesimal generator A of $\{U(t)\}$ restricted to a properly defined domain is skew self-adjoint; $U(t) = \exp(At)$ is then well defined and furnishes us with a weak solution of the wave equation for arbitrarily prescribed initial data of finite energy. The differentiability of these weak solutions for smooth data is proved from known properties of the Laplace operator (which is the essential ingredient of A).

In Section 2 the incoming and outgoing subspaces D_- and D_+ are defined essentially as in the free space problem; more precisely, D_- $[D_+]$ consists of all initial data for which the solution vanishes identically in some spherical neighborhood of the obstacle for all times in the past $t < 0$ [future $t > 0$]. We show that each of these subspaces has the three properties postulated in Chapter II. By far the most difficult of these properties

133

to verify is that the closure of ∪ $U(t)D_{\pm}$ is equal to the space H of all data with finite energy. Our proof of this fact is indirect and uses both spectral theory and harmonic analysis. An easy consequence of this result is that the energy which is contained in any bounded subdomain tends to zero as t tends to infinity for every solution of finite total energy.

In Section 3 we study the associated semigroup $\{Z(t)\}$ and its infinitesimal generator B. The principal result here is that $Z(2\rho)(\kappa I - B)^{-1}$ is a compact operator and according to the general result derived in Chapter III this implies that the spectrum of B, namely $\sigma(B)$, is discrete. In case the obstacle is star-shaped, we show that $Z(t)$ tends in norm to zero as t tends to infinity; from this we conclude that the rate of decay of energy contained in a bounded region is exponential in t, in fact, uniformly so for all signals which originate in a given bounded subdomain. We conjecture on the basis of geometrical optics that such uniform energy decay should be expected if and only if the sojourn time for all reflected rays in some sphere containing the obstacle is bounded.

In Section 4 we conclude from the spectral properties of B and from the connection between A and B that if g has bounded support and μ does not lie in the spectrum of B then the equation $(\mu I - A)f = g$ has a local solution f which is eventually outgoing and analytic in μ. It follows from this that the μ-outgoing Green's function for the reduced wave operator $(\Delta - \mu^2)$ can be continued meromorphically from the left to the right half-plane and that its poles consist of $\sigma(B)$. Furthermore we show that the reduced wave equation has a solution which vanishes on the obstacle and which is eventually μ-outgoing if and only if μ belongs to $\sigma(B)$. Finally we derive the limiting amplitude principle; that is we obtain the asymptotic behavior of solutions of the inhomogeneous wave equation $u_{tt} - \Delta u = \exp(\mu t) g$ for g of bounded support and μ not in $\sigma(B)$.

In Section 5 we express the incoming and outgoing spectral representations for the exterior problem in terms of scattered plane waves and show that the scattering operator is of the form: Identity plus an integral operator whose kernel is the asymptotic value at infinity of the scattered plane waves. The entries of the scattering matrix are just the values of this kernel and these quantities can be measured directly by observations made at large distances from the scattering object. Therefore it is important to be able to deduce properties of the scattering object from those of the scattering matrix. Our only contribution to this problem is a proof that the former is uniquely determined by the latter.

1. The Hilbert Space H and the Group $\{U(t)\}$

In this section we shall study solutions of the wave equation in an exterior domain G which satisfy the boundary condition of being zero on ∂G. We shall show that, given initial data with finite energy there exists a unique solution with this data and, if the data satisfy certain conditions, that the solution is C^∞. Finally we show how to construct solutions for initial data which are only locally of finite energy. We assume throughout this chapter that G is of class C^2 and that ∂G is bounded.

We introduce the following notation: $G(R)$ denotes the set of those points in G for which $|x| < R$; and $E(u(t), R)$ stands for the energy contained in the domain $G(R)$ of a solution u of the wave equation at time t, that is

$$E(u(t), R) = \frac{1}{2} \int_{G(R)} \{|u_x(x, t)|^2 + |u_t(x, t)|^2\}\, dx.$$

We begin by proving the following classical inequality:

Theorem 1.1. *Let $u(x, t)$ be a smooth solution of the wave equation*

$$u_{tt} - \Delta u = 0$$

for all x in G and all t, satisfying the boundary condition

$$u(x, t) = 0 \qquad for \quad x \quad on \quad \partial G.$$

Then the following energy inequality holds:

(1.1) $$E(u(T), R) \leq E(u(0), R + T).$$

For the sake of completeness we sketch the proof:

Multiply the wave equation by u_t and integrate over the space-time region: x in G, $|x| < R + T - t$, $0 < t < T$. Integration by parts yields the integral identity

(1.2) $$\frac{1}{2} \int_{G(R)} \{|u_x(T)|^2 + |u_t(T)|^2\}\, dx$$

$$- \frac{1}{2} \int_{G(R+T)} \{|u_x(0)|^2 + |u_t(0)|^2\}\, dx + \int_{\partial G} u_t u_n\, dS$$

$$+ \frac{1}{2^{3/2}} \int_{|x|=R+T-t} (u_t^2 - 2u_t u_r + u_x^2)\, dS = 0.$$

Because of the boundary condition the boundary integral over ∂G is zero; the boundary integral over the cone is clearly nonnegative, so the energy inequality (1.1) follows from the above identity.

Corollary 1.1. *If the initial data for u vanish in the ball $\{|x| < R\}$, then $u(x, t)$ vanishes in the cone $\{|x| < R - t\}$.*

In particular, if the initial data for u are zero throughout G, it follows that u is zero for all time.

Reversing the direction of time we get the inequality

$$E(u(T), R) \geq E(u(0), R - T),$$

and letting R tend to infinity we obtain

Corollary 1.2. *If initially the total energy of u is finite, u has the same total energy for all time.*

From the *a priori* estimate (1.1) it is not hard, using the methods of partial differential equations, to give a proof of the existence of solutions with prescribed initial data. We shall however give another, operator theoretic, proof here which fits in better with our Hilbert space approach to scattering theory.

We denote by $f = \{f_1, f_2\}$ pairs of complex-valued functions defined in G which shall serve as initial data. We define the *energy norm* as follows:

$$|f|_E^2 = \frac{1}{2} \int_G \{|\partial_x f_1|^2 + |f_2|^2\} \, dx \, ;$$

the Hilbert space H is defined as the completion in the energy norm of smooth data with compact support in G.

For future reference we introduce the following further notation:

H_D is the closure in the Dirichlet norm of smooth scalar valued functions u with compact support in G; the Dirichlet norm is defined as

$$|u|_D^2 = \int_G |\partial_x u|^2 \, dx \quad .$$

So H consists of vectors whose first component belongs to H_D, whose second component is square integrable, and

$$|f|_E^2 = |f_1|_D^2 + |f_2|_0^2$$

where the zero norm denotes the L_2 norm.

Let G' be a subdomain of G; the local energy $|f|_E^{G'}$ is defined by integrating over G' instead of all of G.

Note that H has a natural imbedding as a subspace of H_0 ; this amounts to defining the data to be zero inside $R_n - G$.

Our aim is to construct the one-parameter family of operators $\{U(t)\}$ which assign to given initial data f the data of the corresponding solution of the wave equation at time t. Clearly, these operators $\{U(t)\}$ form a group; further it follows from Corollary 1.2 that $U(t)$ is unitary. Such a one-parameter group of unitary operators (which moreover is strongly continuous) is completely characterized by its infinitesimal generator A. We now proceed in the converse direction and construct the infinitesimal generator A:

Definition.   The operator A is defined as

$$ A = \begin{pmatrix} 0 & I \\ \Delta & 0 \end{pmatrix} ; $$

its domain $D(A)$ is the set of all data $f = \{f_1, f_2\}$ such that Af lies in H. This means that Δf_1, defined in the sense of distributions, is square integrable over G and that f_2 is square integrable and belongs to H_D .

Theorem 1.2.   *A as defined above is skew self-adjoint.*

Proof.   We have to show that $D(A)$ is dense in H and that $A^* = -A$. The former is obvious since $D(A)$ includes all $C_0^\infty(G)$ data.

To show that $A^* = -A$ we turn to the definition of A^*:

$A^* g = h$, with g, h in H, means that for all f in $D(A)$

(1.3) $$ (Af, g)_E = (f, h)_E . $$

We prove first that A is skew symmetric and hence that A^* extends $-A$. For arbitrary f in $D(A)$ and g in $C_0^\infty(G)$ an integration by parts gives

$$ (Af, g)_E = \frac{1}{2} \int_G [\partial_x f_2 \, \overline{\partial_x g_1} - \partial_x f_1 \, \overline{\partial_x g_2}] \, dx . $$

Now for any g in $D(A)$ the second component g_2 belongs to both H_D and L_2, and it follows from this that there exists an approximating sequence $\{g^n\}$ contained in $C_0^\infty(G)$ such that $g^n \to g$ in the H metric and at the same time $g_2^n \to g_2$ in the H_D metric. The above relation therefore holds for all

g in $D(A)$ from which we conclude that

$$(Af, g)_E = -(f, Ag)_E.$$

In order to prove that $-A$ extends A^*, let g be an arbitrary element of $D(A^*)$; then g satisfies (1.3). Taking f to have first component zero we obtain

$$(f_2, g_1)_D = (f_2, h_2)_0.$$

In particular, for f_2 in $C_0^\infty(G)$ if we integrate by parts on the left we obtain

$$-(\Delta f_2, g_1)_0 = (f_2, h_2)_0$$

which shows that

(1.4) $$-\Delta g_1 = h_2$$

in the distribution sense.

Next we choose an f whose second component is zero; (1.3) gives

(1.5) $$(\Delta f_1, g_2)_0 = (f_1, h_1)_D.$$

We choose f_1 the following way: let ϕ be an arbitrary C_0^∞ function. It follows from classical estimates [e.g., inequality (1.2) of Chapter IV] that for any compact subdomain G' of G and for all ψ in H_D

$$\int_{G'} |\psi|^2 \, dx \leq \text{const} \, |\psi|_D^2 \ .$$

Taking G' to be the support of ϕ we obtain from the above that

$$|(\phi, \psi)_0| \leq \text{const} \, |\psi|_D \ ;$$

thus, the linear functional $l(\psi) = \overline{(\phi, \psi)_0}$ is bounded in the Dirichlet norm. Therefore, by the Riesz representation theorem there exists an f_1 in H_D such that

(1.6) $$(f_1, \psi)_D = (\phi, \psi)_0$$

for all ψ in H_D. Taking ψ to be $C_0^\infty(G)$ and integrating by parts, we conclude that

$$-(\Delta f_1, \psi)_0 = (\phi, \psi)_0,$$

so that

(1.7) $$-\Delta f_1 = \phi$$

in the sense of distributions.

Substituting (1.7) into the left side of (1.5) and using (1.6) on the right with $\psi = h_1$, we obtain

$$-(\phi, g_2)_0 = (\phi, h_1)_0$$

which shows that

(1.8) $-g_2 = h_1 .$

Equations (1.4) and (1.8) together show that $g = \{g_1, g_2\}$ belongs to $D(A)$ and that $h = -Ag$. Since $h = A^*g$, this completes the proof of Theorem 1.2.

Remark. Another way of proving Theorem 1.2 is to show as above that A is skew symmetric and then that the deficiency indices of A are zero. The latter amounts to showing that $I \pm A$ maps $D(A)$ onto H and this can be done by making use of the Riesz representation theorem as above to prove that the equation

$$(I \pm A)f = g$$

has a solution f in $D(A)$ for all g such that g_1 is in L_2. The set of all such g is dense in H; and since, as is easily shown, the operator A is closed it follows that $I \pm A$ is indeed surjective.

Having shown that A as defined above is skew self-adjoint it now follows from Stone's theorem (see Appendix 1) that A generates a group of operators $\{U(t)\}$ with the following properties:

(a) $U(t)$ is unitary,
(b) $\{U(t)\}$ forms a one-parameter group,
(c) $U(t)$ is strongly continuous in t,
(d) $U(t)f$ is strongly differentiable with respect to t if and only if f belongs to $D(A)$, in which case

(1.9) $\dfrac{d}{dt} U(t)f = A U(t)f ,$

(e) $U(t)$ maps $D(A)$ onto $D(A)$ and commutes with A.

Suppose that f belongs to $D(A)$ and denote the first component of $U(t)f$ by $u(x, t)$. Then the second component of the relation (1.9) gives

(1.10) $u_{tt} = \Delta u ;$

that is, u satisfies the wave equation in the sense of distributions.

Next we show that for such solutions u the results contained in Theorem 1.1 and its corollaries remain true; that is, the integrations by parts performed in the proof of Theorem 1.1 are also valid for u.

Lemma 1.1. *For f in $D(A)$ the function $u(x, t) = [U(t)f]_1(x)$ has second derivatives which are square integrable with respect to x for every t.*

Proof. Since f belongs to $D(A)$ so does $U(t)f$, and hence by the definition of A the Laplacian of the first component is square integrable. Next we make use of the following estimate from the theory of elliptic equations:

If $u(x)$ *belongs to* H_D *and* Δu *is square integrable, then all second derivatives of u are square integrable, and*

$$(1.11) \qquad | \partial_x^2 u |_0 \leq \text{const} \{| \Delta u |_0 + | u |_D\} \,.$$

This result shows that the pure second x-derivatives of u are square integrable; the square integrability of the mixed x, t derivatives follows from the fact that the first component of $d/dt\, U(t)f$ belongs to H_D ; finally, the square integrability of u_{tt} follows from (1.10). This completes the proof of Lemma 1.1.

We leave it to the reader to verify that if $u(x, t)$ has square integrable second partial derivatives and if u and u_t belong to H_D then the first derivatives of u are square integrable on hypersurfaces, u_t vanishes on ∂G and Green's formula (1.2) is valid. It follows then that Theorem 1.1 holds for u.

Although the proof of Theorem 1.1 makes use of the square integrability of the second derivatives of u, the conclusion (1.1) involves only the first derivatives. Therefore, if u is the limit in the energy norm of a sequence of solutions u_n for which (1.1) is valid, it follows that (1.1) is valid for u also.

Let f be any element of H and $u = [U(t)f]_1$; then by definition of H, f is the limit in the energy norm of a sequence f_n of $C_0^\infty(G)$ data. These belong to $D(A)$ and therefore, as shown above, (1.1) holds for $u_n = [U(t)f_n]_1$. Therefore, by the above observation it holds for u also.† We summarize this result as

Theorem 1.3. *Theorem* 1.1 *and its corollaries hold for all solutions u of the form* $[U(t)f]_1$, f *in H.*

---

† The same limiting procedure shows that u is a weak solution of the wave equation.

We remark that by using estimates from the theory of elliptic equations pertaining to higher derivatives and using Sobolev's lemma one can easily show that if f belongs to $\cap_k D(A^k)$, then $U(t)f$ is C^∞ in x and t. Since we have no need for this result we omit it.

Theorem 1.4. *Let F be the set of data such that*

$$(1.12) \qquad\qquad |Af|_E + |f|_E \leq 1 \; ;$$

then F is precompact in the local energy norm $|f|_E^{G'}$ for any bounded subset G' of G.

Proof. According to the definition of A, the inequality (1.12) can be expressed as

$$(1.12)' \qquad\qquad |f_2|_0 + |\partial_x f_2|_0 + |\partial_x f_1|_0 + |\Delta f_1|_0 \leq 1 \; .$$

Since f_1 belongs to H_D the estimate (1.11) applies and hence

$$|\partial_x^2 f_1|_0 \leq \text{const} \{|\Delta f_1|_0 + |\partial_x f_1|_0\} \; .$$

Combining this with $(1.12)'$ we see that the integrals of the squares of all derivatives of f_1 up to order two, and of f_2 up to order one are uniformly bounded for all f in F. It now follows from Rellich's compactness theorem that for any bounded subset G' of G the set of f_1 are precompact in the norm $|f_1|_D^{G'}$, and the set of f_2 are precompact in the norm $|f_2|_0^{G'}$. This completes the proof of Theorem 1.4.

Data f are called *locally in H* if ϕf belongs to H for every C_0^∞ scalar factor ϕ; f belonging locally to $D(A)$ is defined likewise. We shall now show how to define $U(t)f$ for f locally in H; by a partition of unity we decompose f as

$$f = \sum f_j \, ,$$

where each f_j belongs to H and f_j is zero for all $|x| < j$. We define

$$U(t)f = \sum U(t)f_j \, .$$

It follows from Corollary 1.1 that this sum is locally finite, i.e., for x, t in any given bounded set only a finite number of the terms are nonzero. It is easy to verify that this definition is independent of the particular decomposition employed for f.

2. Energy Decay and Translation Representations

In the present section we shall apply the theory developed in Chapter II to the group $\{U(t)\}$ studied in the previous section. The appropriate incoming and outgoing subspaces can be defined as follows: Choose $\rho > 0$ so that the ball $\{|\,x\,| < \rho\}$ contains ∂G in its interior and set $D_+{}^\rho = U_0(\rho)D_+$ and $D_-{}^\rho = U_0(-\rho)D_-$, where D_+ and D_- denote the outgoing and incoming subspaces of data in free space. Note that data f in $D_+{}^\rho\;[D_-{}^\rho]$ are characterized by the property that $[U(t)f](x)$ vanishes in the truncated forward [backward] cone:

$$|\,x\,| < t + \rho \qquad (t > 0)\,,$$

$$[|\,x\,| < -t + \rho \qquad (t < 0)]\,.$$

Theorem 2.1. $D_+{}^\rho$ *is an outgoing subspace; that is, $D_+{}^\rho$ is a closed subspace satisfying the conditions:*

(i) $U(t)D_+{}^\rho \subset D_+{}^\rho$ *for* $t > 0$,

(ii) $\cap\, U(t)D_+{}^\rho = \{0\}$,

(iii) $\overline{\cup\; U(t)D_+{}^\rho} = H$.

Likewise, $D_-{}^\rho$ is an incoming subspace. Moreover, $D_+{}^\rho$ and $D_-{}^\rho$ are orthogonal.

Proof. That $D_+{}^\rho$ is closed and satisfies the properties (i) and (ii) is proved exactly as in the free space case. This is also true of the orthogonality of $D_+{}^\rho$ and $D_-{}^\rho$; in fact since they are also subspaces of H_0 it follows from Corollary 2.2 of Chapter IV that they are orthogonal in H.

Property (iii) lies considerably deeper and will be proved by means of a series of lemmas. We start with the observation that property (iii) is closely related to the *local decay of energy*, i.e., the following property:

(2.1) $$\lim_{t\to\infty} |\,U(t)f\,|_E{}^{G'} = 0$$

for all f in H and every *bounded* subdomain G' of G. We first show

Lemma 2.1. *Property* (iii) *implies the relation* (2.1).

Proof. If property (iii) holds, then given any f in H and any $\epsilon > 0$ there exists a number T and an element g in $D_+{}^\rho$ such that

$$|\,f - U(T)g\,|_E < \epsilon\,.$$

Let R be chosen large enough so that G' is contained in the ball $\{|x| < R\}$. Then $U(t + T)g$ vanishes in G' for $t > R - T$ and since $|U(t)f - U(t + T)g|_E = |f - U(T)g|_E < \epsilon$, it follows that

$$|U(t)f|_E^{G'} < \epsilon$$

for all $t > R - T$. This together with the fact that ϵ is arbitrary proves Lemma 2.1.

Actually, we are interested in the inverse implication; it turns out that property (iii) even follows from a much weaker assertion of local energy decay, namely

$$(2.1)_w \qquad\qquad \liminf_{t \to \infty} |U(t)f|_E^{G'} = 0$$

for all f in H and all bounded subdomains G' of G.

Lemma 2.2. *The relation* $(2.1)_w$ *implies property* (iii).

Proof. Suppose that property (iii) is false. Then, by the orthogonal projection theorem there exists a nonzero g in H orthogonal to all data of the form $U(t)D_+^\rho$, t positive or negative. This can also be expressed by saying that $U(t)g$ is orthogonal to D_+^ρ for all t.

Next we use the fact that H is imbedded in H_0 so that the free space operator $U_0(s)$ is applicable to elements in H. By definition $U_0(\rho)D_+ = D_+^\rho$. Hence, referring to Theorem 2.3 of Chapter IV we see that the free space translation representation of D_+^ρ is all of $L_2((\rho, \infty) \times S_{n-1})$ and since $U(t)g$ lies in the orthogonal complement of this set we infer that $U_0(-\rho)U(t)g$ belongs to D_-. In other words $[U_0(-s)U(t)g](x)$ *vanishes in the backward cone* $\{|x| < s - \rho\}$ from which it follows for $s > 2\rho$ that $U_0(-s)U(t)g$ is also a solution of the mixed problem; in fact

$$(2.2) \qquad U_0(-(s + 2\rho))U(t)g = U(-s)U_0(-2\rho)U(t)g$$

for $s \geq 0$.

Next, let $G(R)$ denote the ball $\{x; |x| < R\}$. By $(2.1)_w$ we see that given $\epsilon > 0$ and k there exists a $t(k, \epsilon) > (k + 1)\rho$ such that

$$|U(t)g|_E^{G(5\rho)} < \epsilon.$$

Applying Theorem 1.3, which also holds for the free space solution, we have

$$(2.3) \quad |U_0(-2\rho)U(t)g|_E^{G(3\rho)} < \epsilon \qquad \text{and} \qquad |U(t - 2\rho)g|_E^{G(3\rho)} < \epsilon.$$

Since $U_0(-s)U(t)g$ and $U(-s)U(t)g = U(-s+t)g$ have the same Cauchy data outside the ball $\{|x| < \rho\}$ at $s = 0$, these two solutions will be equal in the domain of dependence of this set, namely in the set $\{|x| > |s| + \rho\}$. In particular,

$$[U_0(-2\rho)U(t)g](x) = [U(t-2\rho)g](x)$$

for $|x| > 3\rho$. Thus, the difference of the two sides is zero for $|x| > 3\rho$; combining this fact with (2.3) we obtain

$$|\, U_0(-2\rho)U(t)g - U(t-2\rho)g\,|_E < 2\epsilon\,.$$

Finally, if we apply the unitary operator $U(2\rho - t)$ to this difference and make use of the relation (2.2) with $s = t - 2\rho$ we

$$|\, U_0(-t)U(t)g - g\,|_E < 2\epsilon\,.$$

Recalling that $[U_0(-t)U(t)g](x)$ vanishes for $|x| < t - \rho$ and that $t > (k+1)\rho$, we conclude that

$$|\, g\,|_E^{G(k\rho)} < 2\epsilon\,.$$

Since this holds for any k and any ϵ, it follows that $|\, g\,|_E = 0$, contrary to our supposition that g is nonzero. This proves Lemma 2.2.

It remains to prove that the energy decays; this property turns out to be connected with the spectral properties of the infinitesimal generator A. We show the nature of this connection by establishing the following related assertion:

If the spectrum of A is absolutely continuous, then $U(t)$ tends weakly to zero as t tends to infinity for each f in H.

Proof. We represent $U(t)$ as

$$(2.4) \qquad\qquad (U(t)f, g)_E = \int e^{i\lambda t}\, d(P(\lambda)f, g)_E\,,$$

where $\{P(\lambda)\}$ is the spectral family of projection operators for A. The absolute continuity of the spectrum of A means that for any f and g the scalar measure

$$dm(\lambda) = d(P(\lambda)f, g)_E$$

is absolutely continuous. It then follows from the Riemann–Lebesgue lemma that (2.4), which is the Fourier transform of dm, tends to zero as t tends to infinity, as asserted

This result is used in the second part of Appendix 2 but not in the present chapter.

From the fact that $U(t)$ tends weakly to zero it is not hard to deduce the local energy decay; this will be done below. However, it is not easy to verify (although it follows *a posteriori*) that the operator A has an absolutely continuous spectrum. The following weaker result, due to Rellich, suffices for our purpose and can be deduced directly:

Theorem 2.2.   *The generator A has no point spectrum.*

Our proof of this result will be delayed until after we have established the energy decay by means of the assertion of the theorem.

Lemma 2.3.   *If A has no point spectrum, then there exists a sequence $\{t_k\}$ tending to infinity such that $\{U(t_k)\}$ tends weakly to zero.*

Our proof of this lemma is based† in turn on the following classical result due to Wiener:

Proposition 2.1.   *Let dm be a signed measure of finite total variation and containing no point measure. Then the mean value of the square of its Fourier transform is zero.*

For the sake of completeness we now include a proof of Wiener's theorem: Let $\hat{m}$ denote the Fourier transform of dm; since the total variation of dm is finite we conclude that $\hat{m}$ is bounded. The Fourier transform of $m(\lambda + \epsilon) - m(\lambda - \epsilon)$ is readily computed to be $t^{-1}\sin \epsilon t\, \hat{m}(t)$, a square integrable function; and hence by the Parseval relation we have

$$(2.5) \quad \text{const} \int_{-\infty}^{\infty} \mid m(\lambda + \epsilon) - m(\lambda - \epsilon) \mid^2 d\lambda = \int_{-\infty}^{\infty} \frac{\sin^2 \epsilon t}{t^2} \mid \hat{m}(t) \mid^2 dt.$$

Set

$$M(\epsilon) = \sup_{\lambda} \mid m(\lambda + \epsilon) - m(\lambda - \epsilon) \mid.$$

† We wish to thank Karel deLeeuw who suggested to us this use of Wiener's theorem. Our original proof of the energy decay, presented in Appendix 2, relies on the theory of almost periodic functions.

Then the integral on the left in (2.5) is bounded by

$$M(\epsilon) \int |m(\lambda + \epsilon) - m(\lambda - \epsilon)| \, d\lambda,$$

which can be rewritten as

$$M(\epsilon) \sum_{k=-\infty}^{\infty} \int_{(2k-1)\epsilon}^{(2k+1)\epsilon} |m(\lambda + \epsilon) - m(\lambda - \epsilon)| \, d\lambda$$

$$= M(\epsilon) \int_{-\epsilon}^{\epsilon} \sum_{k=-\infty}^{\infty} |m[(2k+1)\epsilon + \tau] - m[(2k-1)\epsilon + \tau]| \, d\tau$$

$$\leq M(\epsilon) 2\epsilon V,$$

where V is the total variation of dm. Since dm contains no point measure and since it has finite total variation, $m(\lambda)$ is uniformly continuous on the augmented real numbers and this implies that $M(\epsilon)$ tends to zero with ϵ. It follows then from the above estimate that the left side of (2.5) and hence the right side is of $o(\epsilon)$.

Making use of the inequality

$$2\theta/\pi \leq |\sin \theta| \qquad \text{for} \qquad |\theta| \leq \pi/2,$$

we find that the right side of (2.5) is bounded from below by

$$\frac{1}{T^2} \int_{-T}^{T} |\hat{m}(t)|^2 \, dt,$$

where T abbreviates $\pi/2\epsilon$. It now follows from our previous estimate that

(2.6)
$$\frac{1}{T} \int_{-T}^{T} |\hat{m}(t)|^2 \, dt$$

tends to zero as T becomes infinite; this is the assertion of Wiener's theorem.

From the fact that (2.6) tends to zero we infer the

Corollary 2.1.   *Given any two positive numbers c and d, then for T large enough the Lebesgue measure of the set of points t in the interval $(-T, T)$ where $|\hat{m}(t)| \geq d$ is less than cT.*

We now return to the proof of Lemma 2.3. We choose a denumerable dense subset $\{f_i\}$ of H. For each integer k choose $d = 1/k$ and $c < 1/2k^2$.

Then for T large enough and $\geq k$ there exists a $t_k > T/2$ such that

$$(2.7) \qquad\qquad | \, (U(t_k)f_i \, , f_j)_E \, | \, < 1/k$$

for all $i, j \leq k$. Clearly, the $\{t_k\}$ tend to infinity with k, and since the f_i are dense and the $U(t_k)$ are bounded in norm, it follows from (2.7) that the $\{U(t_k)\}$ tend weakly to zero, as asserted in Lemma 2.3.

Lemma 2.4.   *The energy decay*

$$(2.1)_w \qquad\qquad \liminf_{t \to \infty} | \, U(t)f \, |_E^{G'} = 0$$

is valid for all f in H and every bounded subdomain G' of G.

Proof.   Since $| \, U(t)f \, |_E^{G'} \leq | \, U(t)f \, |_E \, = \, | \, f \, |_E \,$, it suffices to prove $(2.1)_w$ for all f in a dense subset of H. We take this subset to be the domain of A. For f in $D(A)$ we have

$$| \, U(t)f \, |_E \, = \, | \, f \, |_E \qquad \text{and} \qquad | \, AU(t)f \, |_E \, = \, | \, Af \, |_E \, .$$

According to Theorem 1.4 this implies that the one-parameter set $\{U(t)f;\ t\ \text{real}\}$ is precompact in the local energy norm $| \cdot |_E^{G'}$ for any bounded sub-domain G'. Therefore, given any sequence $\{t_k\}$ we can select a subsequence (denoted in the same way) such that $\{U(t_k)f\}$ converges in the local energy norm. On the other hand, according to Theorem 2.2 and Lemma 2.3 we can choose the sequence $\{t_k\}$ so that $\{U(t_k)\}$ tends weakly to zero. As a consequence† $\{U(t_k)f\}$ tends to zero in the local energy norm; this completes the proof of Lemma 2.4.

Combining Lemmas 2.4 and 2.2 we conclude that property (iii) in Theorem 2.1 holds. The only step missing in this argument is the proof of Theorem 2.2.

Proof of Theorem 2.2.   Since A is skew self-adjoint, its spectrum is purely imaginary and so we have to prove that σi, σ real, is not an eigenvalue.

The value $\sigma = 0$ has to be treated in a special way which we now do. Suppose that f belongs to $D(A) \subset H$ and satisfies

$$Af = 0 \, .$$

---

† The first derivatives of the first component and the second component of $U(t_k)f$ tend to zero in the local L_2 norm.

For the components f_1, f_2 of f this means that f_2 is zero and that f_1 satisfies the Laplace equation

$$\Delta f_1 = 0$$

and in addition that f_1 vanishes on ∂G. Multiplying by f_1 and integrating by parts over G we conclude that the Dirichlet integral of f_1 over G vanishes so that f_1 is also zero. This proves that $\sigma = 0$ is not an eigenvalue of A.

Next let σ denote a nonzero real number and suppose that f belongs to $D(A) \subset H$ and satisfies the equation

$$(2.8) \qquad Af = i\sigma f .$$

We shall deduce that f is zero without making use of the fact that f vanishes on ∂G. To this end let ϕ be any C^∞ scalar function which vanishes near ∂G and is equal to one for all $|x| > \rho$. Define g to be ϕf in G and equal to zero outside of G. Then g is defined everywhere, has finite energy, and satisfies

$$(2.9) \qquad (A_0 - i\sigma)g = h ;$$

here h vanishes in the region where ϕ is identically one, that is for $|x| > \rho$.

According to Corollary 4.3 of Chapter IV the data g vanishes for $|x| > \rho$; hence, f itself vanishes for $|x| > \rho$. In G the function f satisfies the elliptic equation (2.8); the solutions of such an equation are analytic and therefore it follows that f is zero throughout G, as asserted.

The Rellich uniqueness theorem which follows is a simple consequence of Theorem 2.2.

Theorem 2.3.   *Let f be a local solution of the eigenvalue equation (2.8), $\sigma \neq 0$ and real, in some exterior domain G and suppose that f is eventually outgoing. Then f is identically zero in G.*

Proof.   Since f is eventually outgoing, $[U_0(t)f](x)$ is zero for $|x| < t -$ const; in particular for T large enough $U_0(t)f$ vanishes in the forward cone $\{|x| < \rho + t - T\}$ and

$$(2.10) \qquad U(t) U_0(T)f = U_0(t + T)f$$

for $t > 0$. On the other hand f satisfies the eigenvalue equation (2.8) and it follows from this that

$$(2.11) \qquad U(t)f = e^{i\sigma t}f .$$

Next define

$$f_T = U(T)f - U_0(T)f.$$

By a domain of dependence argument it follows that f_T is zero for $|x| > T + \rho$. Therefore, f_T has finite energy, say E, and the same is true of $U(t)f_T$. In particular, the energy of $U(t)f_T$ inside the ball of radius $t + \rho$ is less than E; that is

(2.12) $| U(t)f_T |_E^{G(t+\rho)} \leq E$.

Now we see from (2.10) and (2.11) that for t positive

(2.13) $U(t)f_T = \exp [i\sigma(t + T)]f - U_0(t + T)f.$

Since f is eventually outgoing the previous choice of T enables us to assert that $U_0(t + T)f$ is zero in $G(t + \rho)$; and thus we conclude from (2.11) that

$$U(t)f_T = \exp [i\sigma(t + T)]f \quad \text{in} \quad G(t + \rho) .$$

Substituting this into (2.12) we see, since σ is real, that f itself has finite energy in $G(t + \rho)$; since t is arbitrary the total energy of f is less than E and hence it follows from Theorem 2.2 that f vanishes in G. This completes the proof of Theorem 2.3.

Remark 2.1. If $\operatorname{Im} \sigma < 0$ then the above proof works even without recourse to Theorem 2.2.

The representation theory developed in Chapter II is applicable to the group $\{U(t)\}$ and the incoming and outgoing subspaces D_{-}^{ρ} and D_{+}^{ρ}. In particular, there exists an outgoing translation representation under which D_{+}^{ρ} maps onto $L_2(0, \infty ; N)$ and by Theorem 3.1 of Chapter II this representation is unique to within an isomorphism on N.

We now show how to relate the outgoing translation representation for $\{U(t)\}$ to the free space translation representation. The thing to notice is that $U(t)$ and $U_0(t)$ act in the same way on D_{+}^{ρ} when t is positive. As we have already remarked, $U_0(\rho)D_{+} = D_{+}^{\rho}$ so that D_{+}^{ρ} maps onto $L_2(\rho, \infty ; N)$ in the free space translation representation. Hence, if f in D_{+}^{ρ} maps onto k_0 in the free space translation representation, then the mapping

$$f \rightarrow k_+(s) = k_0(s + \rho)$$

takes D_{+}^{ρ} onto $L_2(0, \infty ; N)$ in such a way that $U(t)f$ goes into $k_+(s - t)$ for all $t > 0$. If we now simply extend this mapping to all f which are eventually in D_+^ρ so as to preserve this property, then we will obtain an outgoing translation representation for $\cup \, U(t)D_+^\rho$. This mapping is obviously extendable by continuity to the closure of $\cup \, U(t)D_+^\rho$ which by Theorem 2.1 is H.

Suppose that in the above outgoing translation representation the representer k_+ of f is smooth with bounded support; we claim that

$$(2.14) \qquad \lim_{t \to \infty} t^{(n-1)/2} u_t(x \,,\, t) \;=\; k_+(s - \rho \,,\, \theta)$$

where the limit is taken along the ray $x = (t + s)\theta$; here $u_t(x, t) = [U(t)f]_2(x)$. To prove this we note that if $k_+(s)$ has bounded support, then there is a T such that $U(T)f$ belongs to D_+^ρ and has representer $k_+(s - T) = k_0(s - T + \rho)$. Theorem 2.4 of Chapter IV applied to $U_0(t')U(T)f = U(t' + T)f$, with $t' > 0$, yields

$$\lim_{t' \to \infty} (t')^{(n-1)/2} u_t[(s + t')\theta \,,\, t' + T] = k_0(s - T \,,\, \theta)$$

and setting $t' + T = t$ we obtain (2.14).

It is easy at this point to connect our presentation with the usual formalism of scattering theory in which one begins with the wave operators:

$$W_\pm = \text{strong lim}_{t \to \pm \infty} U(-t)U_0(t) \;.$$

In the first place, if f belongs to D_+^ρ then $U(-t)U_0(t)f = f$ for $t > 0$ so that W_+ restricted to D_+^ρ acts like the identity. Similarly, if $U(T)f$ lies in D_+^ρ then $W_+ f = U(-T)U_0(T)f$. Hence, if we consider W_+ as a mapping from the free space translation representation to the outgoing translation representation of $\{U(t)\}$ as defined above, then

$$k_0(s) \to k_+(s) = k_0(s + \rho) \,,$$

at least for representers $k_0(s)$ which vanish say for $s < -T$. Since data with such representers are dense in H_0, W_+ has the above representation for all f in H_0. This not only proves that W_+ exists on all of H_0 but that its range is all of H. Likewise W_- can be realized as the mapping

$$k_0(s) \to k_-(s) = k_0(s - \rho)$$

from the free space translation representation to the incoming translation representation for $\{U(t)\}$.

Since W_+ and W_- are unitary maps on H_0 to all of H, the scattering operator

$$S = W_+^{-1}W_-$$

is well defined and unitary on H_0 to itself. If one makes use of the incoming and outgoing translation representations, then S can be realized as the mapping

$$k_-(s + \rho) \rightarrow k_+(s - \rho)$$

on the free space translation representation to itself; here $k_-(s)$ and $k_+(s)$ are respective incoming and outgoing translation representers of W_-f. This amounts to the definition of S given in Chapter II, aside from an unessential translation of 2ρ.

Remark 2.2. A direct proof, not dependent on the above representation theory, of the basic properties of W_- and W_+ is also at hand. If data f has support in the ball $\{|x| < R\}$, then $U_0(R + \rho)f$ belongs to $D_+{}^\rho$ by Theorem 1.3 of Chapter IV and as above $W_+f = U(-R - \rho)U_0(R + \rho)f$. Hence W_+ is well defined and isometric on a dense subspace in H_0 and can be extended by continuity to all of H_0. A simple calculation shows that W_+ is an intertwining operator for the two groups; that is

$$(2.15) \qquad U(t)W_+ = W_+U_0(t) .$$

It follows from this and the fact that W_+ restricted to $D_+{}^\rho$ acts like the identity, that $U(t)D_+{}^\rho$ belongs to the range of W_+ for all t. We can therefore conclude from Theorem 2.1 that the range of W_+ is all of H. A similar argument applies to W_- .

3. The Semigroup $\{Z(t)\}$

Both the scattering operator and the semigroup $\{Z(t)\}$ which was introduced in Chapter III are determined by the incoming and outgoing subspaces $D_{\mp}{}^\rho$ and because of this these two objects are in many ways the duals of one another. Consequently we can employ the semigroup as a tool in the study of the scattering matrix. In this section we establish some of the properties of the semigroup and conjecture about some others.

We recall the definition of $\{Z(t)\}$:

$$Z^\rho(t) = P_+{}^\rho U(t)P_-{}^\rho \qquad (t \geq 0) ;$$

here $P_-{}^\rho [P_+{}^\rho]$ is the orthogonal projection of H onto the orthogonal complement of D_-^ρ $[D_+^\rho]$. According to Theorem 1.1 of Chapter III the operators $\{Z^\rho(t)\,;\,t \geq 0\}$ form a strongly continuous semigroup of operators on the subspace

$$K^\rho = (D_+^\rho + D_-^\rho)^\perp$$

and they annihilate the subspace $D_+^\rho + D_-^\rho$. We denote the infinitesimal generator of this semigroup by B^ρ. Since the dependence on ρ is of no importance in this section, we omit the superscript for the present; it will however be revived in the next section. We again denote the domain $G \cap \{|x| < k\}$ by $G(k)$.

The following theorem will be used to establish the meromorphic property of the scattering matrix (see Theorem 5.1 of Chapter III).

Theorem 3.1. $Z(2\rho)(\kappa I - B)^{-1}$ is a compact operator for $\kappa > 0$.

The proof of this theorem is based on the next three lemmas which deal with the operator

$$M = U(2\rho) - U_0(2\rho)\,.$$

Lemma 3.1. *The operator M has the following properties:*

(i) $|M| \leq 2$,

(ii)   *For arbitrary f in H, $[Mf](x) = 0$ for $|x| > 3\rho$,*

(iii)

(3.1) $$|Mf|_E \leq 2\,|f|_{E^{G(5\rho)}}.$$

Proof. Since M is the difference of two unitary operators it is clear that $|M| \leq 2$. On the other hand by a domain of dependence argument (Theorem 1.3) we see that $[U(t)f](x) = [U_0(t)](x)$ for $|x| > |t| + \rho$. Thus, $[Mf](x)$ vanishes for $|x| > 3\rho$ and

$$|Mf|_E = |Mf|_{E^{G(3\rho)}} \leq |U(2\rho)f|_{E^{G(3\rho)}} + |U_0(2\rho)f|_{E^{G(3\rho)}}.$$

Another application of the principle of domain of dependence shows that both $|U(2\rho)f|_{E^{G(3\rho)}}$ and $|U_0(2\rho)f|_{E^{G(3\rho)}}$ are bounded by $|f|_{E^{G(5\rho)}}$; this yields inequality (3.1).

Lemma 3.2. *If f is orthogonal to D_-^ρ, then so is $U(t)f$, $U_0(t)f$ for $t > 0$.*

Proof. The subspace D_-^ρ is mapped into itself by $U(-t)$ and $U_0(-t)$ if $t > 0$ and therefore the orthogonal complement of D_-^ρ is mapped into itself by the adjoints of these operators, namely $U(t)$ and $U_0(t)$.

Corollary 3.1. *M maps the orthogonal complement of D_-^ρ into itself.*

Lemma 3.3. $P_+^\rho U_0(2\rho) P_-^\rho = 0.$

Proof. In the free space translation representation, the subspaces D_-^ρ and D_+^ρ correspond to $L_2(-\infty, -\rho; N)$ and $L_2(\rho, \infty; N)$, respectively. The orthogonal complement of D_-^ρ therefore corresponds to $L_2(-\rho, \infty; N)$ and a translation of 2ρ obviously takes this into $L_2(\rho, \infty; N)$; in other words $U_0(2\rho)$ maps the orthogonal complement of D_-^ρ into D_+^ρ and therefore $P_+^\rho U_0(2\rho) P_-^\rho = 0.$

Corollary 3.2. $P_+^\rho U(2\rho) P_-^\rho = P_+^\rho M P_-^\rho.$

Proof of Theorem 3.1. We recall that for $\kappa > 0$

$$(3.2) \qquad (\kappa I - B)^{-1} f = \int_0^\infty e^{-\kappa t} Z(t) f \, dt$$

and

$$(\kappa I - A)^{-1} f = \int_0^\infty e^{-\kappa t} U(t) f \, dt \, ;$$

here A is as before the generator of the group $\{U(t)\}$. Moreover, by Lemma 3.2 $P_-^\rho U(t) P_-^\rho = U(t) P_-^\rho$ so that $P_-^\rho (\kappa I - A)^{-1} P_-^\rho = (\kappa I - A)^{-1} P_-^\rho$. Combining these facts with the Corollary 3.2 we obtain

$$(3.3) \qquad Z(2\rho)(\kappa I - B)^{-1} f = \int_0^\infty e^{-\kappa t} Z(t + 2\rho) f \, dt$$

$$= P_+^\rho U(2\rho) \int_0^\infty e^{-\kappa t} U(t) P_-^\rho f \, dt$$

$$= P_+^\rho U(2\rho) P_-^\rho (\kappa I - A)^{-1} P_-^\rho f$$

$$= P_+^\rho M P_-^\rho (\kappa I - A)^{-1} P_-^\rho f$$

$$= P_+^\rho M (\kappa I - A)^{-1} P_-^\rho f.$$

Next we use the fact that $A(\kappa I - A)^{-1}$ is a bounded operator. More precisely,

$$|A(\kappa I - A)^{-1} P_-^\rho f|_E \le \kappa \, |(\kappa I - A)^{-1} P_-^\rho f|_E + |P_-^\rho f|_E \le 2 \, |f|_E \, .$$

Define the set F as

$$F = \{g\} \, , \qquad g = (\kappa I - A)^{-1} P_-^\rho f \, , \qquad (|f|_E \le 1) \, .$$

According to the inequality above $| Ag |_E \leq 2$ for all g in F. Hence, applying Theorem 1.4 we find that the set F is precompact in the local energy norm $| \cdot |_E^{G(5\rho)}$. If we now make use of the inequality (3.1) we see that MF and hence $P_+^\rho MF$ is precompact in the energy norm. Since by formula (3.3) the set $P_+^\rho MF$ is the image of the unit sphere by the operator $Z(2\rho) \times (\kappa I - B)^{-1}$, it follows that $Z(2\rho)(\kappa I - B)^{-1}$ is a compact operator.

The above theorem together with Corollaries 4.2 and 4.5 of Chapter III give

Corollary 3.3. *The generator B has a pure point spectrum and the resolvent of B is meromorphic in the plane and holomorphic on the imaginary axis.*

As explained at the end of this section, it is unlikely that for highly indented obstacles the energy decay assertion of Lemma 2.1 or the compactness statement of Theorem 3.1 can be greatly improved. However, if one restricts the geometry of the reflecting object, then it is to be expected that much more can be said. The first result of this kind was given by Morawetz [1, 2] who showed for *star-shaped obstacles* and data f with support in the domain $G(k)$ that

$$(3.4) \qquad | U(t)f |_E^{G(k)} \leq \frac{\text{const}}{t} | f |_E .$$

We reproduce this result in Appendix 3. By combining (3.4) with the semigroup properties of $\{Z(t)\}$ Lax *et al.* [1] proved that the energy in fact decays exponentially.

Theorem 3.2. *If the obstacle is star-shaped, then for all data with support in $G(\rho)$ there exist positive constants c and α such that*

$$(3.5) \qquad | U(t)f |_E^{G(\rho)} \leq ce^{-\alpha t} | f |_E \qquad (t > 0) .$$

Proof. Applying Morawetz's estimate (3.4) for $k = 5\rho$ we see that for T sufficiently large

$$(3.6) \qquad | U(T)g |_E^{G(5\rho)} \leq \tfrac{1}{8} | g |_E$$

for all g which has its support in $G(5\rho)$. Now by Lemma 3.1 the data $g = Mf$ has its support in $G(3\rho)$ for any f in H, hence making use of (3.1) and (3.6) we obtain

$$(3.7) \quad | MU(T)Mf |_E \leq 2 | U(T)Mf |_E^{G(5\rho)} \leq \tfrac{1}{4} | Mf |_E \leq \tfrac{1}{2} | f |_E .$$

For $t \geq 4\rho$, Lemmas 3.2 and 3.3 imply that

(3.8) $$Z(t)f = P_{+}{}^{\rho}MU(t - 4\rho)MP_{-}{}^{\rho}f,$$

and combining (3.7) and (3.8) we have

$$| Z(T + 4\rho)f |_{E} \leq \tfrac{1}{2} | f |_{E}.$$

Hence, for any $t > 0$ with $k(T + 4\rho) \leq t < (k + 1)(T + 4\rho)$ we may write

(3.9) $| Z(t)f |_{E} \leq | [Z(T + 4\rho)]^{k}f |_{E} \leq (\tfrac{1}{2})^{k} | f |_{E} \leq ce^{-\alpha t} | f |_{E}$

for some positive constants c and α, and all data f in H.

We note that the data in $D_{\pm}{}^{\rho}$ have their support outside of the ball $\{| x | < \rho\}$. Consequently, data f with support in the ball $\{| x | < \rho\}$ satisfy $P_{-}{}^{\rho}f = f$ and for all data g, $[P_{+}{}^{\rho}g](x) = g(x)$ for $| x | < \rho$. Thus, for f with support in the ball $\{| x | < \rho\}$ it follows from (3.9) that

$$| U(t)f |_{E^{G(\rho)}} = | Z(t)f |_{E^{G(\rho)}} \leq | Z(t)f |_{E} \leq ce^{-\alpha t} | f |_{E},$$

as asserted in Theorem 3.2.

From the exponential decay (3.9) and the Laplace transform representation (3.2) of the resolvent of B we deduce

Corollary 3.4. *If the obstacle is star-shaped, then for some positive constant α the spectrum of B lies in the half-plane* Re $\kappa \leq -\alpha$.

We close this section with some heuristic remarks connecting the geometry of the obstacle and the analytic properties of the semigroup of operators $\{Z(t)\}$; the connecting link, not surprisingly, is geometrical optics. The relevant geometric property seems to us to be the following: Starting at some point x on the sphere $| x | = \rho$ draw a ray in some inward direction ω; if this ray hits the obstacle, reflect it according to the classical laws of reflection. Continuing this process the broken ray will eventually return to the sphere $| x | = \rho$. Let $l(x, \omega)$ denote its total length (possibly infinity) and let $l(G)$ denote the supremum of the $l(x, \omega)$ for all x, ω. Of course, $l(G)$ may be infinite.

Conjecture. (a) If $l(G) < \infty$, then $Z(t)$ is a compact operator for all $t > l(G) + 12\rho$. (b) If $l(G) = \infty$, then $| Z(t) |_{E} = 1$ for all t.

Part (b) of the conjecture is probably not too difficult to prove. Given T one would have to construct a solution of the wave equation almost all of whose energy is contained inside the ball $\{|\,x\,| \leq \rho\}$ for all t less than T A natural way to construct such "energy retaining" solutions is by geometrical optics. The geometrical optics approximations can be restricted to stay in an arbitrarily small neighborhood of any given ray; if $l(G) = \infty$ there exist rays of arbitrary large length which stay inside the sphere $|\,x\,| = \rho$ so that there are geometrical optics approximations which retain energy for an arbitrarily long time. These approximations satisfy an inhomogeneous wave equation whose right hand side is small if the frequency is large; therefore they can be turned into exact solutions by subtracting a small correction term.

The difficulty in this approach is caused by caustics along rays; the analytical form of the geometrical optics solution changes across caustics and it is not easy to construct a uniform formula which will nearly satisfy the wave equation.

Part (a) of the conjecture would follow from the validity of the "generalized Huygens' principle" which is true for solutions of the pure initial value problem and which we conjecture to be true for solutions of the mixed problem. Very roughly, this principle states that "sharp" propagation of signals takes place only along rays, including reflected rays. The proof of this for the pure initial value problem relies on the geometrical optics approximation; there are substantial difficulties in carrying out such a proof for the mixed problem, caused by the presence of "glancing" rays (i.e., rays at a shadow boundary).

We now state two precise and roughly equivalent forms of the generalized Huygens' principle; for the first version we need a representation of the operator $U(t)$ as an integral operator with a distribution kernel R:

$$[U(t)f](x) = \int R(x, y, t)f(y)\, dy \ .$$

(A) $R(x, y, t)$ *is a smooth function at all points* x, y, t *such that* (x, t) *cannot be connected to* $(y, 0)$ *by a reflected ray.*

To state the second version we consider two subdomains G_1 and G_2 such that at time T the subdomain G_2 does not lie in the sharp domain of influence of G_1 in the following sense: no ray originating in $\bar{G}_1$ at time $t = 0$ lies in $\bar{G}_2$ at time T.

(B) *The set $F = \{g\}$, $g = U(T)f$, f supported in G_1, $|f|_E \leq 1$ is pre-compact in the local energy norm $|g|_E^{G_2}$.*

Conjecture (a) follows from (B) by taking G_1 to be $G(3\rho)$, G_2 to be $G(5\rho)$. For $T > l(G) + 8\rho$, G_1 and G_2 have the required property. Using form (3.8) for $Z(t)$ and the properties of M stated in Lemma 3.1 it follows from (B) that for $t = T + 4\rho$, $Z(t)$ maps the unit sphere in H into a pre-compact set. This shows that $Z(t)$ is compact for $t > l(G) + 12\rho$, as asserted.

If conjecture (a) were true we could apply theorems from Chapter III to draw valuable conclusions about the spectrum of B and the asymptotic behavior of $\{Z(t)\}$. It would follow from Theorem 5.3 of Chapter III that only a finite number of points of the spectrum of B lie to the right of any line Re $\sigma = $ const; it would follow from Theorem 5.4 of Chapter III that for large t, $Z(t)$ has an asymptotic expansion in terms of exponentials. Finally, using Corollary 5.1 of Chapter III we could conclude the exponential decay of $Z(t)$ without making use of the estimate (3.4).

The above remarks would apply in particular to star-shaped obstacles as the following proposition shows:

Proposition 3.1. *If the obstacle is star-shaped, then $l(G) \leq 2\rho$.*

Proof. The star-shaped property requires that the outer normal n at a point x on the obstacle satisfies the condition $x \cdot n \geq 0$. Now, if a ray with direction ω hits such an obstacle at x, then the reflected ray has the direction

$$\omega' = \omega_t - (\omega \cdot n)n ,$$

where ω_t denotes the tangential component of ω, and since $\omega \cdot n \leq 0$ we have
$$\omega' \cdot x - \omega \cdot x = -2(\omega \cdot n)(n \cdot x) \geq 0 .$$

It follows from this that the distance from the origin for the undeflected path is always less than the distance from the origin to the corresponding point on the reflected path: that is for all $t \geq 0$

$$|x + \omega t|^2 = x^2 + t^2 + 2x \cdot \omega t \leq x^2 + t^2 + 2x \cdot \omega' t = |x + \omega' t|^2 .$$

Since the undeflected path has a sojourn time in the ball $\{|x| \leq \rho\}$ of at most 2ρ, we see by applying the above argument inductively to the successive reflections of the ray that the sojourn time of the reflected ray is also at most 2ρ.

In case $l(G) = \infty$, conjecture (b) offers no clue about the spectrum of B; we suspect that in this case there is an infinite sequence of eigenvalues μ_n such that $\operatorname{Re} \mu_n \to 0$ and $|\operatorname{Im} \mu_n| \to \infty$. The corresponding eigenfunctions must be connected in some intimate way with the geometry of the obstacle.

4.   The Relation between the Semigroup $\{Z(t)\}$ and the Solutions of the Reduced Wave Equation

As we have seen in Chapter III the spectrum of the semigroup generator B determines the poles of the scattering matrix and hence a characterization of this spectrum is very desirable. In this section we shall prove that the spectrum of B consists of all complex numbers μ for which there is a nontrivial μ-outgoing solution (see Definition 4.2 of Chapter IV) of the reduced wave equation

$$\Delta u - \mu^2 u = 0$$

(4.1)

$$u = 0 \quad \text{on} \quad \partial G .$$

Combining this result with Corollary 3.1 we see that the reduced wave equation (4.1) has only a discrete set of nontrivial μ-outgoing solutions. We are able to go a step further and show, again by means of the semigroup $\{Z(t)\}$, that the Fredholm alternative applies; that is, aside from the above discrete set of exceptional points, the nonhomogeneous equation (4.11) has a unique μ-outgoing solution for right members with compact support.

Theorem 4.1.   *The generator B has μ as an eigenvalue if and only if the equation*

(4.2) $$Af = \mu f$$

has a nontrivial eventually outgoing local solution.

Proof.   For each $a \geq \rho$ we define the semigroup of operators

(4.3) $$Z^a(t) = P_+{}^a U(t) P_-{}^a , \quad \text{for} \quad t \geq 0,$$

where, as before, $P_-{}^a$ $[P_+{}^a]$ is the orthogonal projection of H onto the orthogonal complement of $D_-{}^a$ $[D_+{}^a]$. Again, let B^a denote the infinitesimal generator of $\{Z^a(t)\}$. According to Theorems 6.3 and 6.4 of Chapter III the B^a's have the same spectra and corresponding eigenspaces have the same dimension; as shown there for $a < b$ the operation

(4.4) $$f^a = P_+{}^a f^b$$

maps the null space of $\mu I - B^b$ onto that of $\mu I - B^a$ in a one-to-one fashion. Thus to each eigenpair $\{\mu, f^\rho\}$ of B^ρ and to each $b \geq \rho$ there corresponds an eigenvector f^b such that

$$B^b f^b = \mu f^b \qquad \text{and} \qquad f^\rho = P_+ f^b .$$

Moreover, these eigenvectors satisfy the relation (4.4) whenever $a < b$. Now, for any g, $P^a g = g$ for $|x| < a$, and hence it follows from (4.4) that $f^a(x) = f^b(x)$ for $|x| < a < b$; this shows that the limit

(4.5) $$f(x) \equiv \lim_{a \to \infty} f^a(x)$$

exists for all x in G.

It is clear from (4.5) that f belongs to H locally. However, more is true: f belongs to $D(A)$ locally. In order to prove this we shall employ the following lemma.

Lemma 4.1. *To each e^a in $D(B^a)$ there corresponds a g^a in $D(A)$ such that for $|x| < a$*

$$e^a(x) = g^a(x) \qquad and \qquad [B^a e^a](x) = [A g^a](x) .$$

Proof of Lemma 4.1. According to the relation (3.2)

$$(\lambda I - B^a)^{-1} h = P_+{}^a (\lambda I - A)^{-1} h$$

for h in K^a and $\operatorname{Re} \lambda > 0$. Hence, choosing $h = (\lambda I - B^a) e^a$ and setting $g^a = (\lambda I - A)^{-1} h$ we see that $e^a = P_+{}^a g^a$, from which the first assertion follows. On the other hand $\lambda g^a - A g^a = h = \lambda e^a - B^a e^a$ and this together with the first assertion establishes the second assertion.

Returning to the proof of the theorem, suppose that μ is an eigenvalue of B^ρ; define f^a and f as above. For any ζ in $C_0^\infty(R_n)$ with support in the ball $\{|x| < a\}$

$$\zeta f = \zeta f^a = \zeta g^a$$

where g^a defined as in Lemma 4.1 corresponds to f^a in $D(B^a)$. This shows that f belongs to $D(A)$ locally. Likewise the lemma shows for $|x| < a$ that

$$[Af](x) = [Ag^a](x) = [B^a f^a](x) = \mu f^a(x) = \mu f(x),$$

from which (4.2) follows.

To conclude this part of the argument we note by Lemma 6.1 of Chapter III that f^a is orthogonal to $D_-{}^\rho$. Thus $[U_0(t)f^a](x) = 0$ for $|x| < t - \rho$ and a domain of dependence argument shows that $[U_0(t)f](x) = [U_0(t)f^a](x)$ for $|x| < a - t$. Since a is arbitrary we can assert that

$$[U_0(t)f](x) = 0 \qquad \text{for all} \qquad |x| < t - \rho,$$

and hence that f is eventually outgoing.

To prove the converse, suppose that f is locally in $D(A)$, satisfies (4.2) and is eventually outgoing. Then $\exp(\mu t)f$ is a solution of the mixed problem and hence by uniqueness (Corollary 1.1) $U(t)f = \exp(\mu t)f$. Choosing ζ in $C^\infty(R_n)$ to be identically one for $|x| > \rho$ and identically zero near ∂G, we see that $f' \equiv \zeta f$ belongs to $D(A_0)$ locally and, since we have modified f only on a bounded set, that f' is also eventually outgoing. Clearly,

$$A_0 f' - \mu f' = g,$$

where g is of finite energy and vanishes for $|x| > \rho$.

Applying Theorem 4.1 of Chapter IV we can assert that f' has a free space translation representation k' given by (4.6)' of Chapter IV. Thus k' is locally square integrable and

$$k'(s, \omega) = \begin{cases} 0 & \text{for} \quad s < -\rho, \\[2mm] e^{-\mu s} n(\omega) & \text{for} \quad s > \rho; \end{cases}$$

and it follows from either (4.6)' of Chapter IV or Corollary 4.1 of Chapter IV that $n(\omega)$ lies in $L_2(S_{n-1})$. Since f differs from f' only by data in H with support in the ball $\{|x| < \rho\}$, we see by property (4.2a) of Chapter IV that f too has a free space representation k with the above listed properties.

We now define

$$k^a(s) = \begin{cases} k(s) & \text{for} \quad s < a, \\[2mm] 0 & \text{for} \quad s > a; \end{cases}$$

and let f^a denote the data with representer k^a, that is $f^a = \mathcal{I} k^a$ in the notation of Chapter IV. It is clear that $f^a(x) = f(x)$ for $|x| < a$ so that f^a belongs to H. We recall that $D_-{}^a [D_+{}^a]$ corresponds to $L_2(-\infty, -a; N) [L_2(a, \infty; N)]$ in the free space translation representation (essentially Theorem 2.3 of Chapter IV) and it follows from this that f^a belongs to K^a. All that remains to be shown is that f^a is an eigenfunction of Z^a, i.e., that

$$(4.6) \qquad\qquad Z^a(t)f^a = e^{\mu t}f^a .$$

Since

$$k(s) - k^a(s) = \begin{cases} 0 & \text{for } s < a, \\ e^{-\mu s}n & \text{for } s > a, \end{cases}$$

it follows that $f - f^a$ is outgoing. In fact $U_0(t)(f - f^a)$ vanishes for $|x| < a + t$ and

$$U_0(t)(f - f^a) = U(t)(f - f^a) = e^{\mu t}f - U(t)f^a .$$

Therefore,

$$e^{\mu t}k(s) - k(s - t) + k^a(s - t) = \begin{cases} e^{\mu t}k(s) & \text{for } s < a + t, \\ 0 & \text{for } s > a + t, \end{cases}$$

is a free space translation representation of $U(t)f^a$. This representer of $U(t)f^a$ has finite norm. It follows from this and the fact that $D_+{}^a$ corresponds to $L_2(a, \infty; N)$ that $P_+{}^a U(t)f^a$ is represented by $\exp(\mu t)k^a$. This establishes (4.6) and concludes the proof of Theorem 4.1.

Suppose that $u(x)$ is a μ-outgoing solution of the reduced wave equation (4.1) satisfying the boundary condition in the sense that u belongs to $H_D(G)$ locally and satisfies the reduced wave equation in the sense of distributions. Then $f = \{u, \mu u\}$ belongs to $D(A)$ locally, $Af = \mu f$, and f is eventually outgoing. Combining the above result with the fact that the spectrum of B is discrete (Corollary 3.3) we can state the following extension of the Rellich uniqueness Theorem 2.3:

Corollary 4.1. *Aside from a discrete set of μ's with $\mathrm{Re}\,\mu < 0$, there are no nontrivial μ-outgoing solutions of the reduced wave equation (4.1) satisfying the boundary condition.*

Next we solve the nonhomogeneous problem using the same techniques as above.

Theorem 4.2. *If μ belongs to the resolvent set of the generator B and if h belongs to H and has bounded support (or merely belongs to K^c for some $c > \rho$), then there exists a unique eventually-outgoing local solution of*

$$(4.7) \qquad\qquad \mu f - Af = h \; ;$$

this solution, f_μ, is an analytic function of μ in the strong sense of the local energy norm.

Proof. The uniqueness follows directly from Theorem 4.1. We therefore proceed to the construction of a solution to the Eq. (4.7). Suppose that h belongs to K^c. We note first of all that for all $a > c$

$$f_\mu{}^a \equiv (\mu I - B^a)^{-1}h$$

exists and is analytic for μ in the resolvent set of B^a; by Theorem 6.3 of Chapter III the resolvent set is independent of a and by Corollary 3.3 of the present chapter it is connected. Now according to the relation (3.2) if $a \geq c$ and $\operatorname{Re}\lambda > 0$, then

$$(4.8) \qquad\qquad (\lambda I - B^a)^{-1}h = P_+{}^a \int_0^\infty e^{-\lambda t}U(t)h\,dt\,.$$

Consequently, if $\operatorname{Re}\lambda > 0$, then $f_\lambda{}^a = P_+{}^a f_\lambda{}^b$ for $a < b$ and since by Lemma 3.2 $U(t)h$ is orthogonal to $D_-{}^c$ it follows that $f_\lambda{}^a$ is also orthogonal to $D_-{}^c$. Both of these properties are preserved under analytic continuation and hence we can assert for any μ in the resolvent set of B that

$$(4.9) \qquad\qquad f_\mu{}^a = P_+{}^a f_\mu{}^b \qquad \text{for} \quad a < b$$

and that $f_\mu{}^a$ is orthogonal to $D_-{}^c$. In particular $f_\mu{}^a(x) = f_\mu{}^b(x)$ for $|x| < a$ so that

$$(4.10) \qquad\qquad f_\mu \equiv \lim_{a\to\infty} f_\mu{}^a$$

is well defined and belongs to H locally. As in the proof of Theorem 4.1, we apply Lemma 4.1 to show that f_μ belongs to $D(A)$ locally and that for $|x| < a \; (a > c)$

$$[Af_\mu](x) = [B^a f_\mu{}^a](x) = \mu f_\mu{}^a(x) - h(x) = \mu f_\mu(x) - h(x)\,,$$

from which (4.7) follows.

Since $f_\mu{}^a$ is orthogonal to $D_-{}^c$ we know that $[U_0(t)f_\mu{}^a](x)$ vanishes for $|x| < t - c$. A domain of dependence argument shows that

$$[U_0(t)f_\mu](x) = [U_0(t)f_\mu{}^a](x) \qquad \text{for} \quad |x| < a - t,$$

and a being arbitrary we may conclude that $[U_0(t)f_\mu](x) = 0$ for $|x| < t - c$ and hence that f_μ is eventually outgoing. Finally, the analyticity of $f_\mu{}^a$ for all a proves directly that f_μ is analytic in the local energy norm. This concludes the proof of Theorem 4.2.

Remark 4.1. If we solve (4.7) for a family of h's, all with support in the ball $\{|x| < c\}$, and if h_τ depends analytically on τ, then for $a > c$

$$f_{\mu,\tau}{}^a = (\mu I - B^a)^{-1}h_\tau$$

is clearly analytic in both μ and τ. Consequently, $f_{\mu,\tau} = \lim_{a \to \infty} f_{\mu,\tau}{}^a$ will be analytic in both μ and τ in the local energy norm.

To obtain a μ-outgoing solution to the inhomogeneous reduced wave equation:

$$(4.11) \qquad\qquad \Delta u - \mu^2 u = h, \qquad u = 0 \qquad \text{on} \quad \partial G$$

where h has compact support, we take the eventually outgoing solution of

$$\mu f - A f = \{0, -h\}$$

provided by Theorem 4.2 and set $u = f_1$. Combining this with Corollary 4.1 and Corollary 3.3 we obtain:

Corollary 4.2. *Aside from a discrete set of μ's with* Re $\mu < 0$ *for which* (4.1) *has a nontrivial μ-outgoing solution, there always exists a unique μ-outgoing solution of the inhomogeneous reduced wave equation* (4.11) *satisfying the boundary condition. This solution depends analytically on μ in the local H_D norm.*

Remark 4.2. All of the above results remain valid if we replace outgoing by incoming. This amounts to a reversal in time in which the roles of $D_-{}^a$ and $D_+{}^a$ are interchanged. The resulting semigroup

$$Z_-{}^a(t) = P_-{}^a U(-t) P_+{}^a, \qquad (t \geq 0)$$

is simply the adjoint semigroup to $\{Z^a(t)\}$ and its generator $B_-{}^a$ is $(B^a)^*$.

With these conventions one can read off the proofs for the incoming assertions from the corresponding outgoing proofs.

Remark 4.3. If f and g belong to H and are zero for $|x| > a$ (or merely belong to K^a) then $(Z^a(t)f, g)_E = (U(t)f, g)_E$ and it follows from the relation (3.2) that

$$((\lambda I - B^a)^{-1}f, g)_E = ((\lambda I - A)^{-1}f, g)_E$$

for Re $\lambda > 0$. Now the left member has an analytic continuation throughout the resolvent set of B^a and hence the same is true of the right member *in spite of the fact that the spectrum of A consists of the entire imaginary axis.* Suppose next that $f^a \neq 0$ belongs to the null space of $(\mu_0 I - B^a)$. Then $((\mu I - B^a)^{-1}f^a, f^a)_E$ has a pole at μ_0 and so does the analytic continuation of $((\lambda I - A)^{-1}f^a, f^a)_E$. The assertion about $((\lambda I - A)^{-1}f, g)_E$ remains true for all values of a and in particular it holds for f, g of compact support. Since we can approximate f^a by data with compact support we see that we can choose f and g so that the analytic continuation of $((\lambda I - A)^{-1}f, g)_E$ has a pole at each point of the spectrum of B^a. This furnishes us with another way of characterizing the spectrum of B.

For Re $\lambda > 0$ the operator $(\lambda I - A)^{-1}$ is an integral operator whose kernel is Green's function $G(\lambda; x, y)$. The analytic properties of $((\lambda I - A)^{-1}f, g)$ for f, g with compact support derived above show

Theorem 4.3.   *For fixed $x \neq y$, $G(\lambda; x, y)$ has a meromorphic continuation into the left λ-half plane and has poles on the spectrum of B.*

Let μ be any complex number not in the spectrum of B and g any function of class $L_2(G)$ with bounded support. Let u be a solution in G of the inhomogeneous wave equation

$$(4.12) \qquad u_{tt} - \Delta u = e^{\mu t}g,$$

which is zero on ∂G and whose initial values have finite energy. The following result, known as the *principle of limiting amplitude*, describes the asymptotic behavior of $u(x, t)$ for large t:

Theorem 4.4.   *As t tends to infinity, $u(x, t)$ converges in the local energy norm to the steady state solution $e^{\mu t}v$ where v is the μ-outgoing solution of the reduced inhomogeneous wave equation*

$$(4.13) \qquad \Delta v - \mu^2 v = -g.$$

Proof. As remarked in the derivation of Corollary 4.2 from Theorem 4.2 the μ-outgoing solution v of (4.13) is the first component of f, the eventually outgoing solution of the inhomogeneous equation

$$\mu f - Af = h = (0, g) ;$$

such a local solution f exists according to Theorem 4.2. The difference $u - e^{\mu t}v$ satisfies the homogeneous wave equation while the difference of their Cauchy data, namely

$$d(t) = \{u, u_t\} - e^{\mu t}f$$

satisfy

$$d(t) = U(t) d_0 ,$$

where $d_0 = u_0 - f$.

Our aim is to prove that $d(t)$ tends to zero in the local energy norm. Now the initial data u_0 of $u(t)$ were assumed to belong to H and hence by the energy decay (Lemma 2.1) it follows that $U(t)u_0$ decays locally. Hence it suffices to prove that $U(t)f$ also decays locally.

By construction f is eventually outgoing; therefore, f has a free space translation representer $k(s, \omega)$ which is zero for $s < -\rho$ and exponential in s for $s > \rho$. It is clear from Eq. (4.6)' of Chapter IV that k is locally square integrable. We now decompose k as $k_1 + k_2$, where k_1 is zero for $s < \rho$ and k_2 has compact support. Set $f_1 = \mathcal{J}k_1$, $f_2 = \mathcal{J}k_2$; then $f = f_1 + f_2$ where $U(t)f_1 = U_0(t)f_1$ vanishes for $|x| < t + \rho$ and f_2 has finite energy. Since $f_2 = f - f_1$, we see that f_2 also belongs to H. Thus,

$$U(t)f = U(t)f_1 + U(t)f_2 ;$$

as above $U(t)f_1$ vanishes for $|x| < t + \rho$ and since f_2 is of finite energy, Lemma 2.1 implies that $U(t)f_2$ tends to zero in the local energy norm. This completes the proof of Theorem 4.4.

5. The Scattering Matrix

Much of the material in this book culminates in the present section. In particular we will obtain basic information about the scattering matrix by means of the theory developed in Chapters II and III joined with the properties of the wave equation already established.

As previously noted the representation theory of Chapter II is applicable since the solution to the exterior problem generates a group of unitary operators $\{U(t)\}$ for which $D_{-}{}^{\rho}$ and $D_{+}{}^{\rho}$ play the role of incoming and outgoing subspaces. The auxiliary space N which appears in this theory was identified in section two as $L_2(S_{n-1})$. This space is clearly separable and hence according to Theorem 4.1 of Chapter II the scattering operator $\mathbb{S}$ on $L_2(-\infty, \infty; N)$ can be realized as a multiplicative operator-valued function $\mathbb{S}(\sigma)$ on N into N having the properties:

(a) $\mathbb{S}(\sigma)$ is the boundary value of an operator-valued function $\mathbb{S}(z)$ analytic for $\operatorname{Im} z < 0$ which converges strongly along the lines $\operatorname{Re} z = \sigma$ to $\mathbb{S}(\sigma)$ for almost all σ,

(b) $|\mathbb{S}(z)| \leq 1$ for all z with $\operatorname{Im} z < 0$,

(c) $\mathbb{S}(\sigma)$ is unitary for almost all σ.

Moreover, the operator $Z(2\rho)(\lambda I - B^a)^{-1}$ is compact by Theorem 3.1. Consequently, we can invoke Theorem 5.1 of Chapter III and combining this with Theorems 3.1 and 4.1 we obtain:

Theorem 5.1.   *The scattering matrix* $\mathbb{S}(z)$ *is holomorphic on the real axis and meromorphic in the whole plane, having a pole at exactly those points* z *for which there is a nontrivial eventually outgoing local solution of*

$$Af = izf.$$

Likewise Theorem 5.2 of Chapter III and the expression (3.9) yield:

Theorem 5.2.   *If the obstacle is star-shaped, then for some positive constant* α *the scattering matrix* $\mathbb{S}(z)$ *is holomorphic and bounded in the strip* $\operatorname{Im} z < \alpha$.

It is also easy to obtain from our development a description of the incoming and outgoing spectral representations of the group $\{U(t)\}$ in terms of incoming and outgoing solutions of the reduced wave equation, and it turns out that the asymptotic values of these solutions furnish the kernel of the scattering matrix $\mathbb{S}(\sigma)$. We shall denote by $\tilde{f}, \tilde{f}_{-}, \tilde{f}_{+}$ the free space, incoming, and outgoing spectral representers, respectively, of a given initial data f. These spectral representers are given by scalar products of f with certain improper eigenfunctions of A_0 and A, respectively. In the case of $\{U_0(t)\}$ these improper eigenfunctions consist of exponential "plane waves," whereas for $\{U(t)\}$ they consist of exponentials plus initially incoming or eventually outgoing terms. Thus the free space spectral representation as established in Theorem 2.1 of Chapter IV can be stated as follows:

The mapping

(5.1) $$f \rightarrow \tilde{f}(\sigma, \omega) = (f, \phi(\cdot, \sigma, \omega))_E$$

where

$$\phi(x, \sigma, \omega) = \frac{(i\sigma)^{(n-3)/2}}{(2\pi)^{n/2}} \{\exp(-i\sigma x\omega), i\sigma \exp(-i\sigma x\omega)\},$$

defines a unitary spectral representation for $\{U_0(t)\}$ *of* H_0 *onto* $L_2(-\infty, \infty, N)$; *here* $N = L_2(S_{n-1})$.

We recall that by Theorem 2.3 of Chapter IV this spectral representation for $\{U_0(t)\}$ is simultaneously incoming and outgoing so that the incoming and outgoing subspaces D_-^0 and D_+^0 for $\{U_0(t)\}$ are mapped onto $A_-(N)$ and $A_+(N)$, respectively. Consequently, the incoming and outgoing subspaces for $\{U(t)\}$, D_-^ρ and D_+^ρ, are mapped onto $\exp(-i\sigma\rho)A_-(N)$ and $\exp(i\sigma\rho)A_+(N)$, respectively in the free space spectral representation, whereas they are mapped onto $A_-(N)$ and $A_+(N)$ in the respective incoming and outgoing spectral representations of $\{U(t)\}$ constructed according to the general theory of Chapter II. If we introduce the slight modification in these representations of multiplying the incoming [outgoing] representers by $\exp(-i\sigma\rho)$ [respectively $\exp(i\sigma\rho)$] then the representers of data in D_-^ρ [D_+^ρ] will coincide with the corresponding free space representers.

Theorem 5.3. *For real* σ *let* v_+ [v_-] *be the* $i\sigma$-*outgoing* [$i\sigma$-*incoming*] *solution of the reduced wave equation*

(5.2) $$\Delta v + \sigma^2 v = 0 \qquad in \quad G$$

which satisfies the boundary condition

$$v = -\exp(-i\sigma x\omega) \qquad on \quad \partial G.$$

Set

$$\psi_\pm(x, \sigma, \omega) = \frac{(i\sigma)^{(n-3)/2}}{(2\pi)^{n/2}} \{v_\pm(x, \sigma, \omega), i\sigma v_\pm(x, \sigma, \omega)\}$$

and define

$$\phi_\pm = \phi + \psi_\pm.$$

Then the mapping

(5.3) $$f \rightarrow \tilde{f}_\pm(\sigma, \omega) = (f, \phi_\mp(\cdot, \sigma, \omega))_E$$

determines a unitary incoming [outgoing] spectral representation† for $\{U(t)\}$ of H onto $L_2(-\infty, \infty; N)$. For data in $D_-{}^\rho$ $[D_+{}^\rho]$ the incoming [outgoing] representer is the same as in the free space spectral representation.

Remark 5.1. The outgoing and incoming solutions $v_\pm$ are readily obtainable from Corollary 4.2 as follows: Let ζ be an infinitely differentiable scalar function which vanishes identically for $|x| > \rho$ and is identically one in the complement of G; and set $g = -\zeta \exp(-i\sigma x\omega)$ and $h = \Delta g + \sigma^2 g$. If u_+ is the $i\sigma$-outgoing solution of (4.11) satisfying the boundary condition $u_+ = 0$ on ∂G, then

$$v_+ = g - u_+$$

is the desired solution of (5.2). Moreover, the proof of Theorem 4.2 shows that $[U_0(t)\{u_+, i\sigma u_+\}](x)$ vanishes for $|x| < t - \rho$; by Huygens' principle $[U_0(t)\{g, i\sigma g\}](x)$ also has this property and hence so does $[U_0(t)\psi_+](x)$. Finally, we note that g and h are obviously analytic in σ, ω and hence by Corollary 4.2 and Remark 4.1 so is u_+; consequently, $\phi_+(\cdot, \sigma, \omega)$ is analytic in σ, ω in the local energy norm. A similar argument applies to v_- and ϕ_-.

Proof of the Outgoing Spectral Formula. It suffices to prove Theorem 5.3 for a dense subset of C^∞ data with bounded support. If f has bounded support, then $\tilde{f}_+(\sigma, \omega)$ is well defined by formula (5.3) and since $\phi_-(\cdot, \sigma, \omega)$ is continuous in σ, ω it too is continuous in σ, ω. It is also clear that for such data the mapping (5.3) is linear.

We first verify for all f in $D_+{}^\rho \cap C_0^\infty$ (which according to Corollary 2.4 of Chapter IV are dense in $D_+{}^\rho$) that the outgoing and free space spectral representations are the same. Since these representations are given by formulas (5.3) and (5.1), respectively, we have to show that their difference is zero; that is that

$$(f, \psi_-)_E = 0.$$

By assumption f belongs to $D_+{}^\rho$ so that its free space translation representer vanishes for $s < \rho$; by construction ψ_- is eventually incoming and its free

---

† Note that the formula for the incoming representer f_- involves the outgoing ϕ_+, and vica versa.

space translation representer is zero for $s > \rho$. Hence it follows from formula (3.1f) of Chapter IV that f and ψ_- are orthogonal.

Next consider f in $C_0^\infty(G)$ and set $f(t) = U(t)f$. Obviously, f lies in $D(A)$ and for each t the data $f(t)$ have bounded support and lie in $D(A)$. Let $\tilde{f}_+(t)$ denote the representer of $f(t)$. Making use of the differential equation and the fact that ϕ_- satisfies the boundary conditions, a simple integration by parts shows that

$$d\tilde{f}_+(t)/dt = (AU(t)f, \phi_-)_E = -(U(t)f, A\phi_-)_E = i\sigma\tilde{f}_+(t).$$

As a consequence

$$(5.4) \qquad\qquad \tilde{f}_+(t) = e^{i\sigma t}\tilde{f}_+.$$

It follows from this that if the map $f(t) \to \tilde{f}_+(t)$ is an isometry for one value of t then it is so for all values of t.

Suppose that f is of the form $U(-T)g$ where g belongs to $D_+^\rho \cap C_0^\infty$. According to Corollary 2.4 of Chapter IV and Theorem 2.1 of the present chapter the set of all such f is dense in H. Since $f(T) = U(T)f = g$ lies in $D_+^\rho \cap C_0^\infty$ it follows from what was already shown that $\tilde{f}_+(T)$ equals the free space spectral representation of $f(T)$ which is known to be isometric. Because of this and the fact that $U(T)$ is unitary we can conclude directly from formula (5.4) that

$$|f|_E = |\tilde{f}_+|.$$

Extending the mapping by continuity, (5.4) now proves that the resulting map is indeed a spectral representation.

It remains only to show that this mapping is onto. But this follows from the fact that in the free space spectral representation D_+^ρ maps onto $e^{i\sigma\rho}A_+(N)$ and hence onto this same set in the outgoing spectral representation. Thus $U(t)D_+^\rho$ maps onto $\exp[i\sigma(t + \rho)]A_+(N)$ and since these sets are dense in $L_2(-\infty, \infty; N)$ the mapping (5.3) is necessarily onto. This concludes the proof of Theorem 5.3.

The asymptotic behavior for large $|x|$ of an $i\sigma$-incoming solution of the reduced wave equation is given by formula (4.21)' of Chapter IV. According to that formula

$$v_-(r\theta, \omega, \sigma) \sim \frac{e^{i\sigma r}}{r^{(n-1)/2}} s(\theta, \omega; \sigma), \qquad \text{as} \quad r \to \infty.$$

and $s(\theta, \omega; \sigma)$ is analytic in θ, ω, σ.

Theorem 5.4. *The scattering matrix is of the form: identity plus an integral operator with kernel const $s(-\theta, \omega; \sigma)$, that is*

$$(5.5) \quad \tilde{f}_+(\sigma, \omega) = [S(\sigma)\tilde{f}_-(\sigma, \cdot)](\omega)$$

$$= \tilde{f}_-(\sigma, \omega) + \left(\frac{i\sigma}{2\pi}\right)^{(n-1)/2} \int_{|\theta|=1} s(-\theta, \omega; \sigma)\tilde{f}_-(\sigma, \theta) \, d\theta.$$

Proof. We begin with the assumption that $S(\sigma)$ can be expressed as the identity plus an integral operator with kernel $K(\theta, \omega; \sigma)$. In this case the relation $\tilde{f}_+ = S\tilde{f}_-$ takes the form

$$(f, \phi_-(\cdot, \sigma, \omega))_E$$

$$= (f, \phi_+(\cdot, \sigma, \omega))_E + \int (f, \phi_+(\cdot, \sigma, \theta))_E K(\theta, \omega; \sigma) \, d\theta.$$

According to Remark 5.1 the improper eigenfunctions $\phi_+(\cdot, \sigma, \theta)$ are continuous in σ, θ in the local energy norm so that we can bring the integral under the energy inner product, at least for f with bounded support. Consequently, in order to justify our assumption it suffices to show with the above choice of K that

$$(5.6) \quad \phi(x, \sigma, \omega) \equiv \phi_-(x, \sigma, \omega) - \phi_+(x, \sigma, \omega) - \int \phi_+(x, \sigma, \theta)\overline{K(\theta, \omega; \sigma)} \, d\theta$$

vanishes identically. Now the integrated term in (5.6) being a superposition of local solutions of $Af = i\sigma f$ for $|x| > \rho$, is itself a local solution. Hence ϕ is a local solution of $A\phi = i\sigma\phi$; we shall show that with the above choice of K, ϕ is eventually outgoing and hence the Rellich uniqueness theorem (Theorem 2.3) implies that ϕ is identically zero.

With this in mind we compute a free space translation representer of ϕ and choose K so that this representer is zero for $s < -\rho$. It is readily verified from (2.7)–(2.9) of Chapter IV that

$$[\Re\phi_0(\cdot, \sigma, \omega)](s, \theta) = \frac{(-1)^{(n-3)/2}}{(2\pi)^{1/2}} e^{-i\sigma s} \delta(\theta - \omega) .$$

Further, Corollary 4.1 of Chapter IV shows that for $s < -\rho$

$$[\Re\psi_-(\cdot, \sigma, \omega)](s, \theta) = e^{-i\sigma s} n(\theta; \sigma, \omega) ,$$

where, by Theorem 4.5 of Chapter IV and the material immediately preceding it, the function $n(\theta; \sigma, \omega)$ is analytic in θ, ω, σ and

$$(5.7) \quad v_-(r\theta, \sigma, \omega) \sim \frac{-(2\pi)^{n/2}}{(i\sigma)^{(n-1)/2}} \frac{e^{i\sigma r}}{r^{(n-1)/2}} n(-\theta; \sigma, \omega) \qquad \text{as} \quad r \to \infty .$$

The translation representer of $\psi_+(x, \sigma, \omega)$ vanishes for $s < -\rho$. Thus we see that the free space translation representer of ϕ for $s < -\rho$ is

$$n(\theta; \sigma, \omega) - \frac{(-1)^{(n-3)/2}}{(2\pi)^{1/2}} \int \delta(\theta' - \theta) \overline{K(\theta', \omega; \sigma)} \, d\theta'$$

times the factor $\exp(-i\sigma s)$; clearly this is zero if

$$\overline{K(\theta, \omega; \sigma)} = (-1)^{(n-3)/2}(2\pi)^{1/2} n(\theta; \sigma, \omega) .$$

Combining this with (5.7) we obtain (5.5).

Theorem 5.5. *The scattering matrix satisfies the following relations:*

$$(5.8) \quad s(-\omega, \theta; \sigma) + (-1)^{(n-1)/2} s(-\theta, \omega; \sigma)$$

$$= -\left(\frac{i\sigma}{2\pi}\right)^{(n-1)/2} \int s(-\omega, \theta'; \sigma) \overline{s(-\theta, \theta'; \sigma)} \, d\theta'$$

$$= -\left(\frac{i\sigma}{2\pi}\right)^{(n-1)/2} \int \overline{s(-\theta', \omega; \sigma)} s(-\theta', \theta; \sigma) \, d\theta' ;$$

$$(5.9) \qquad\qquad s(\theta, \omega; -\sigma) = \overline{s(\omega, \theta; \sigma)} \quad ;$$

$$(5.10) \qquad\qquad s(\theta, \omega; -\sigma) = \overline{s(\theta, \omega; \sigma)} \quad ;$$

$$(5.11) \qquad\qquad s(\theta, \omega; \sigma) = s(\omega, \theta; \sigma) \quad .$$

Remark 5.2. Property (5.11) is known as the *reciprocity law*; it is an immediate consequence of the relations (5.9) and (5.10) which express respectively the time reversibility and the reality of the governing equation. It follows from (5.10) that the singularities of $S(z)$ are symmetrically placed with respect to the imaginary axis.

Proof of Theorem 5.5. The relation (5.8) is a restatement of the unitary property of the scattering matrix, namely,

$$S^*(\sigma)S(\sigma) = I = S(\sigma)S^*(\sigma) .$$

The relation (5.9) is obtained from the behavior of the group $\{U(t)\}$ under the time reversal operator T:

$$T\{f_1, f_2\} = \{f_1, -f_2\}.$$

It is readily verified that $T = T^*$, $T^2 = 1$, and further that

$$TA_0 = -A_0 T \quad \text{and} \quad TA = -AT;$$

it therefore follows that

$$TU_0(t) = U_0(-t)T \quad \text{and} \quad TU(t) = U(-t)T$$

Consequently, the wave operators $W_\pm$ are transformed as $W_\pm T = TW_\pm$ and since $S = W_+^{-1}W_-$ we see that

(5.12) $$ST = TS^{-1}.$$

Now T interchanges incoming and outgoing solutions. Thus

$$T\psi_+ = \frac{(i\sigma)^{(n-3)/2}}{(2\pi)^{n/2}} \{v_+(x, \sigma, \omega), -i\sigma v_+(x, \sigma, \omega)\}$$

is an incoming solution satisfying the same boundary condition as $v_-(x, -\sigma, -\omega)$. Since replacing σ by $-\sigma$ also gives the desired factor in the second component, it follows that

$$T\phi_\pm(x, \sigma, \omega) = (-1)^{(n-3)/2}\phi_\mp(x, -\sigma, -\omega).$$

As a consequence

$$[\widetilde{Tf}]_\pm(\sigma, \omega) = (Tf, \phi_\mp(\cdot, \sigma, \omega))_E = (f, T\phi_\mp(\cdot, , \omega))_E$$
$$= (-1)^{(n-3)/2}\tilde{f}_\mp(-\sigma, -\omega).$$

Combining this with (5.12) and the fact that $S^{-1}(\sigma) = S^*(\sigma)$, we obtain

$$[\widetilde{STf}]_+(\sigma, \omega) = S(\sigma)[\widetilde{Tf}]_-(\sigma, \omega)$$

$$= \tilde{f}_+(-\sigma, -\omega) + \left(\frac{i\sigma}{2\pi}\right)^{(n-1)/2}\int \overline{s(-\theta, \omega; \sigma)}\tilde{f}_+(-\sigma, -\theta)\, d\theta;$$

$$[\widetilde{TS^{-1}f}]_+(\sigma, \omega) = [\widetilde{S^{-1}f}]_-(-\sigma, -\omega)$$

$$= \tilde{f}_+(-\sigma, -\omega) + \left(\frac{i\sigma}{2\pi}\right)^{(n-1)/2}\int s(\omega, \theta, -\sigma)\tilde{f}_+(-\sigma, \theta)\, d\theta;$$

from which (5.9) follows.

The relation (5.10) comes from the fact that A_0 and A are real operators; in terms of the conjugation operator $C\{f_1, f_2\} = \{\bar{f}_1, \bar{f}_2\}$ this means that $CA_0 = A_0 C$ and $CA = AC$. Consequently, $CU_0(t) = U_0(t)C$, $CU(t) = U(t)C$, and

$$(5.13) \qquad\qquad\qquad SC = CS$$

$$C\psi_\pm = \frac{(-i\sigma)^{(n-3)/2}}{(2\pi)^{n/2}} \{\overline{v_-(x, \sigma, \omega)}, \overline{-i\sigma v_-(x, \sigma, \omega)}\}$$

is outgoing and incoming with $\psi_\pm$; and further $v_\pm(x, \sigma, \omega)$ satisfies the same boundary condition as $v_\pm(x, -\sigma, \omega)$. Arguing as above we see that

$$[C\phi_\pm](\sigma, \omega) = \phi_\pm(x, -\sigma, \omega)$$

Consequently,

$$[\widetilde{Cf}]_\pm(\sigma, \omega) = (Cf, \phi_\mp(\cdot, \sigma, \omega))_E = \overline{(f, [C\phi_\mp](\cdot, \sigma, \omega))_E}$$
$$= \overline{\tilde{f}_\pm(-\sigma, \omega)}$$

Inserting this into (5.13) we obtain (5.10).

As a practical consideration it is of interest to know whether or not the asymptotic behavior of the solution to the exterior problem as expressed by the scattering operator uniquely determines the scattering object. This is the uniqueness question in the so-called inverse scattering problem. In the case of the Schrödinger wave equation a solution to this problem has been obtained only for the case of a spherically symmetric potential (see Gelfand and Levitan [1]). In our case the boundedness of the scatterer and the analyticity of the differential operator combine to give the following simple solution to the inverse problem.

Theorem 5.6. *The scattering operator uniquely determines the scatterer.*

Remark. We shall present two proofs; the first which is along classical lines and is due to M. Schiffer (personal communication); whereas the second uses techniques developed in this monograph.

Schiffer's proof. We consider two exterior problems; the symbols for the second will be primed. This proof begins with the assumption that the kernels of the integral operators entering in the scattering operators are the same; that is we assume that $s(\theta, \omega, \sigma) \equiv s'(\theta, \omega, \sigma)$. According to Theorem 5.4

$$v_-(r\theta, \sigma, \omega) - v_-'(r\theta, \sigma, \omega) = O\left(\frac{1}{r^{(n+1)/2}}\right) \qquad \text{as} \quad r \to \infty \quad,$$

and it follows from this, Theorem 4.5 of Chapter IV and Corollary 4.3 of Chapter IV that $\psi \equiv \psi_- - \psi_-'$ vanishes for $|x|$ sufficiently large and hence it follows by analyticity that $\phi_-(x, \sigma, \omega) = \phi_-'(x, \sigma, \omega)$ on $G \cap G'$. Since either ϕ_- or ϕ_-' vanishes at each point of $\partial(G \cap G')$, we conclude that both ϕ_- and ϕ_-' vanish at all points of $\partial(G \cap G')$.

Suppose next that $G'' = G - G \cap G'$ is nonempty. Then $\phi_-(x, \sigma, \omega)$ vanishes on $\partial G''$ and satisfies $A\phi_- = i\sigma\phi_-$ inside of G''; that is ϕ_- restricted to G'' is an eigenvector of the generator A'' with domain G''. However, it is well known that the operator A'' corresponding to a bounded domain can have only a denumerable number of eigenvalues; thus, aside from this exceptional set of σ's, $\phi_-(x, \sigma, \omega)$ must vanish on G'' and hence by analytic continuation throughout G. However, this is impossible since ϕ_- behaves asymptotically like $\exp(i\sigma x\omega)$ for large $|x|$. We conclude that $G = G'$.

The Second Proof of Theorem 5.6. In this proof we simply assume that $S = S'$. Then $W_+^{-1}W_- = W_+'^{-1}W_-'$ and we can set

$$W = W_-'W_-^{-1} = W_+'W_+^{-1}.$$

As is readily verified W is unitary from H onto H' and satisfies the properties:

(i) W restricted to $D_-^\rho + D_+^\rho$ is the identity,
(ii) $WU(t) = U'(t)W$.

We first show that for any f in H, $[Wf](x) = f(x)$ for $|x| > \rho$. Take any g in D_-^ρ, then by (i) above $Wg = g$ and thus by (ii)

(5.14) $U'(t)g = WU(t)g$.

Next we decompose $U(t)g = f_1 + f_2$ and $U'(t)g = f_1' + f_2'$ into orthogonal summands f_1, f_1' in $D_-^\rho + D_+^\rho$ and f_2, f_2' in the orthogonal complement K^ρ of this subspace. Then by (i) $Wf_1 = f_1$ and since W is unitary Wf_2 belongs to K^ρ; hence by (5.14) we see that $f_1 = Wf_1 = f_1'$ and $Wf_2 = f_2'$. On the other hand by a domain of dependence argument $[U(t)g](x) =$

$[U'(t)g](x)$ for $|x| > |t| + \rho$, and since $f_1 = f_1'$ we conclude that $f_2(x) = f_2'(x)$ for $|x| > |t| + \rho$. Considered as data in H_0, f_2 and f_2' have translation representers with support in $|s| < \rho$ so that by Theorem 3.3 of Chapter IV the spherical harmonic coefficients of f_2 and f_2' are analytic in $r = |x|$ for $r > \rho$. Since corresponding coefficients were already shown to be equal for $r > |t| + \rho$ it follows that they are equal for all $r > \rho$; consequently, $f_2(x) = f_2'(x)$ for $|x| > \rho$. This proves that $[Wf](x) = f(x)$ for $|x| > \rho$ and all f of the form $U(t)g$, g in $D_{-\rho}$. According to Theorem 2.1 the union of $U(t)$ images of $D_{-\rho}$ is dense in H and therefore it follows that $[Wf](x) = f(x)$ for all $|x| > \rho$ and all f in H.

We recall that the time reversal operator T introduced in the proof of Theorem 5.5 satisfies $W_+T = TW_-$ and $W_+'T = TW_-'$; it follows from this that $WT = TW$. In particular, then,

$$W\{f_1, 0\} = \tfrac{1}{2}W(f + Tf) = \tfrac{1}{2}(Wf + TWf) = \{[Wf]_1, 0\} ;$$

thus, if f has second component zero, so does Wf. Now choose $f = \{f_1, 0\}$ in H so that f_1 is harmonic in $|x| < 2\rho$ and identically one on $|x| = 2\rho$; according to the classical Dirichlet principle f is of minimum energy as compared with all g in H which are equal to f for $|x| > 2\rho$. Since W is unitary and since $[Wf](x) = f(x)$ for $|x| > 2\rho$ it follows that Wf is of minimum energy as compared with all g in H' which are equal to $f(x)$ for $|x| > 2\rho$. It therefore follows by the Dirichlet principle that $\{f_1', 0\} = Wf$ is also harmonic for $|x| < 2\rho$. But we have shown above that $f_1(x) = f_1'(x)$ for $|x| > \rho$ so that by analyticity we can conclude that $f_1(x) = f_1'(x)$ on $G \cap G'$. Since either f_1 or f_1' vanishes at each point of $\partial(G \cap G')$ we see that both functions vanish at all points of $\partial(G \cap G')$. On the other hand f_1 [and f_1'] is greater than zero on $G \cap \{|x| < 2\rho\}$ [respectively $G' \cap \{|x| < 2\rho\}$] and hence we can conclude that $\partial G = \partial G'$.

6. Notes and Remarks

The material in Section 1 on the solution of the wave equation in an exterior domain is essentially classical; however, as far as we know the infinitesimal generator has never been displayed so explicitly before. The first result on the local energy decay for solutions of the wave equation was obtained by Morawetz [1] for star-shaped obstacles. Our original proof for arbitrary smooth bounded obstacles (reproduced in Appendix 2) appeared in Lax and Phillips [2], and combining our methods with those of Morawetz the three of us were able to prove the exponential decay for star-shaped obstacles (Lax *et al.* [1]). The Rellich uniqueness theorem plays a central role in our proof of the energy decay theorem and one of the consequences of our approach is a new proof of this result. The local energy decay of solutions to the wave equation is also closely connected with the fact that the ranges of the incoming and outgoing wave operators are equal; this result has also been obtained by Birman [1] and Kato [1] who study the wave equation problem by means of properties of the corresponding Schrödinger wave operator.

The existence of outgoing solutions to the reduced wave equation for μ imaginary goes back to Weyl [1], Müller [1], Kupradse [1], and more recently to Werner [1] and Mizohata [1]. Our result is the first which treats the problem for arbitrary complex μ although von Schwarze [1] has shown independently that the Fredholm alternative applies in this case. In the same way our proof of the limiting amplitude principle extends the earlier work of Ladyzhenskaya [1], Eidus [1], Buchal [1], and Morawetz [2, 3].

The meromorphic properties of the scattering matrix are also new although van Kampen [1] previously showed for a centrally symmetric potential of bounded support that the scattering matrix has a holomorphic extension into the lower half-plane. The connection between the scattering operator and the scattering matrix derived in section five has been found in the case of the Schrödinger operator by Ikebe [1]. We note that an operator inner factor such as the scattering operator which differs from the identity by an integral operator can probably be studied with the aid of infinite determinants. Our proof of the uniqueness for the inverse scattering problem is new.

CHAPTER VI

Symmetric Hyperbolic Systems, the Acoustic Equation with an Indefinite Energy Form, and the Schrödinger Equation

In the present chapter we apply our general theory of scattering to symmetric hyperbolic systems for which all sound speeds are different from zero and to the acoustic equation with a potential which can cause the energy form to be indefinite. The scattering operator for the acoustic equation with a potential will be related to the scattering operator for the Schrödinger equation.

It comes as a pleasant surprise that these new applications can be easily treated within the framework of our previous theory. Some changes have to be made, to be sure, but they are merely variations on a theme. In the case of symmetric systems, the multiplicity of sound speeds complicates our analysis of the free space problem but little else is affected. For the acoustic equation with an indefinite energy form the natural incoming and outgoing subspaces are no longer orthogonal; however, they are almost orthogonal in the sense that each contains a subspace with finite relative codimension and these subspaces are orthogonal. This introduces a finite number of poles in the lower half-plane for the scattering matrix.

PART 1

SYMMETRIC HYPERBOLIC SYSTEMS

1. Translation Representation in Free Space

Just as in our treatment of the acoustic equation, we start with a study of the solutions of a symmetric hyperbolic system of equations with constant coefficients in free space. The work in this section parallels and leans on Chapter IV.

The equations which we consider are of the form

$$(1.1) \qquad u_t = A_0 u = \sum_{i=1}^{n} a^i u_{x^i} \; ;$$

$u(x, t)$ is a vector-valued function whose values lie in k-dimensional complex Euclidean space C_k, and the a^i are $k \times k$ Hermitian symmetric matrices with the following property:

For every real nonzero point ξ in R_n

$$(1.2) \qquad a(\xi) \equiv \sum_{i=1}^{n} a^i \xi^i$$

is nonsingular.† This requirement means that the sound speeds, which are the eigenvalues of $a(\xi)$, are all different from zero. Since scattering theory deals with the asymptotic behavior of solutions at infinity, this requirement is almost essential.‡

We require, as before, that the number n of space dimensions be *odd*.

The initial value problem for the system (1.1) is to find solutions with prescribed initial values:

$$u(x, 0) = f(x) \; .$$

---

† Suppose that $n > 1$; then the set of nonzero ξ's is connected. If the condition (1.2) is satisfied, then for all such ξ the matrix $a(\xi)$ does not have zero as an eigenvalue and therefore $a(\xi)$ has the same number of positive eigenvalues for all nonzero ξ's. In particular $a(\xi)$ and $a(-\xi) = -a(\xi)$ have the same number of positive eigenvalues which proves that $a(\xi)$ has as many positive as negative eigenvalues. Therefore the order k of a^i is even. More generally it can be shown (see Adams *et al.* [1]) that (1.2) can hold if and only if $n \le 2\nu + 1$, where ν is the largest power of 2 dividing k.

‡ More precisely, even if the system does have zero sound speeds, scattering theory for such a system is concerned only with solutions for which the modes propagating with zero speed are not excited.

Theorem 1.1. (a)   *The initial value problem for* (1.1) *has a unique* C^∞ *solution* $u(x, t)$ *for all* C_0^∞ *initial data* f. *We shall denote this solution by* $u = Wf$. *The "energy," that is the square integral*

$$\int |u(x, t)|^2 dx,$$

is independent of t. (b)   *The energy of a solution* u *contained at time* s *inside the ball* $\{|x| < R - c_{\max}s\}$ *does not exceed the energy of* u *at time zero contained inside the ball* $\{|x| < R\}$. (c)   *If* $f(x) = 0$ *outside the ball* $\{|x| < R\}$, *then the solution* $u = Wf$ *is zero inside the cones* $\{|x| < c_{\min}|t| - R\}$.

Here $c_{\max}$ *and* $c_{\min}$ *are positive constants whose values are given explicitly in the relation* (1.11).

Proof.   Given a vector-valued function f of class C_0^∞ we shall a little further on construct by Fourier transformation a C^∞ solution with initial data f. That this solution conserves energy follows from (b) and this in turn shows that such solutions are uniquely determined by their initial data.

Now (b) can be deduced by the standard energy method: We form the scalar product of (1.1) with u and take its real part:

$$(1.3) \qquad \mathrm{Re}\, u \cdot u_t - \sum_{i=1}^{n} \mathrm{Re}\, u \cdot a^i u_{x^i} = 0 \,.$$

Next we note that all of the terms are perfect derivatives; that is

$$\mathrm{Re}\, u \cdot u_t = \tfrac{1}{2}(u \cdot u)_t \,,$$

and, because of the Hermitian symmetry of a^i,

$$\mathrm{Re}\, u \cdot a^i u_{x^i} = \tfrac{1}{2}(u \cdot a^i u)_{x^i} \,.$$

Consequently the integral of (1.3) over a domain G in x, t space is equal to the following surface integral over the boundary ∂G of G:

$$0 = \frac{1}{2} \int_{\partial G} \left[u \cdot u n_t - \sum (u \cdot a^i u) n_i \right] dS$$

$$(1.4) \qquad = \frac{1}{2} \int_{\partial G} u \cdot (n_t I - \sum a^i n_i) u \, dS \,;$$

here $\{n_i, n_t\}$ are the components of the outer normal to ∂G. We now choose

G to be the truncated cone:

$$|x| \le R - ct \qquad (0 \le t \le s) .$$

The surface of G consists of three pieces: top, bottom, and mantle. The integral over the top and bottom is

$$\frac{1}{2} \int_{|x| \le R - cs} |u^2(x, s)| \, dx - \frac{1}{2} \int_{|x| \le R} |u^2(x, 0)| \, dx ,$$

which is the difference between the energy contained inside the ball $\{|x| \le R - cs\}$ at time s and the energy contained inside the ball $\{|x| \le R\}$ at time 0; this difference is the quantity asserted to be non-positive in part (b). According to the identity (1.4) this difference is the negative of the integral over the mantle:

$$- \int_{\text{mantle}} u \cdot (n_t I - \sum a^i n_i) u \, dS ;$$

and this expression is negative if the integrand is positive. On the mantle n_t is proportional to c and n_i to $x_i/|x|$ and hence the integrand will be positive if c is greater than the supremum of the eigenvalues of $a(\omega)$ for $|\omega| = 1$. We denoted this quantity by $c_{\max}$.

It follows from (b) that if the initial value of u is zero inside the ball $\{|x| < R\}$ then $u(x, t)$ is zero inside the cones: $\{|x| < R - c_{\max}|t|\}$. This expresses the fact that what happens at time $t = 0$ outside the ball $\{|x| < R\}$ does not influence what happens at time t inside the ball $\{|x| < R - c_{\max}|t|\}$; in other words *signals do not propagate with speed greater than* $c_{\max}$.

On the other hand part (c) asserts that *signals do not propagate with speed less than* $c_{\min}$. This is a kind of Huygens' principle for equation of the form (1.1); its proof will be given later in this section. The constant $c_{\min}$ turns out to be the infinum of the positive eigenvalues of $a(\omega)$ for $|\omega| = 1$.

Let $U_0(t)$ denote the operator relating the initial value of solutions of (1.1) to their value at time t. Using the norm preserving property of these solutions we can extend the operator $U_0(t)$ to all data f in the Hilbert space H_0 of square integrable C_k-valued functions with norm

$$|f| = \left\{ \int |f(x)|^2 \, dx \right\}^{1/2} .$$

The operators $\{U_0(t)\}$ so extended define a one-parameter group of unitary operators on H_0.

We now study the group $\{U_0(t)\}$ by solving equation (1.1) explicitly with the aid of the Fourier transformation: Define

$$(1.5) \qquad \hat{u}(\xi, t) = \frac{1}{(2\pi)^{n/2}} \int e^{i\xi x} u(x, t) \, dx \; ;$$

equation (1.1) then becomes

$$(1.6) \qquad \hat{u}_t = -ia(\xi)\hat{u} \; .$$

In order to solve this vector differential equation we decompose it according to the eigenvectors of $a(\xi)$. We denote the normalized eigenvectors by $r_j(\xi)$, and the corresponding eigenvalues by $\tau_j(\xi)$:

$$(1.7) \qquad -a(\xi)r_j = \tau_j r_j \; .$$

Finally let ϕ_j denote the r_j-component of $\hat{u}$:

$$\phi_j(\xi, t) = \hat{u}(\xi, t) \cdot r_j(\xi).$$

Taking the scalar product of (1.6) with r_j we get, using (1.7) and the symmetry of $a(\xi)$, that

$$\partial_t \phi_j = i\tau_j \phi_j \; .$$

The solution of this scalar differential equation is

$$\phi_j(\xi, t) = \phi_j(\xi) \exp\left[i\tau_j(\xi)t\right],$$

where $\phi_j(\xi)$ is the jth component of the Fourier transform of the initial data f:

$$(1.8) \qquad \phi_j(\xi) = \hat{f}(\xi) \cdot r_j(\xi) \; .$$

We can now combine these components to obtain $\hat{u}$:

$$(1.9) \qquad \hat{u}(\xi, t) = \sum_{j=1}^{k} \phi_j(\xi) \exp\left[i\tau_j(\xi)t\right] r_j(\xi).$$

We label the eigenvalues $\{\tau_j\}$ in decreasing order:

$$\tau_1(\xi) \geq \tau_2(\xi) \geq \cdots \geq \tau_k(\xi).$$

Since $a(\xi)$ is an odd function of ξ, it follows that

(1.10) $\tau_j(-\xi) = -\tau_{j'}(\xi)$

where j' stands for $(k - j + 1)$. Further it is clear that $\tau_j(\xi)$ is positive-homogeneous of degree one. Since we have assumed in (1.2) that none of the τ_j are zero, it follows by the argument given in the footnote (†) on page 178 that exactly one half of the τ_j are positive, the other half negative. We recall that we denoted the infinum of the positive eigenvalues by $c_{\min}$ and the supremum by $c_{\max}$:

(1.11) $c_{\min} = \min_{|\xi|=1} \tau_{k/2}(\xi)$ and $c_{\max} = \max_{|\xi|=1} \tau_1(\xi)$.

The normalized eigenvectors $r_j(\xi)$ are determined up to a numerical factor of absolute value one; in case of multiple eigenvalues the choice is even wider. We choose this numerical factor so that

(1.12) $r_j(\xi) = r_{j'}(-\xi)$ $(j' = k - j + 1)$.

We require the $r_j(\xi)$ to be merely measurable functions of ξ; since continuity is not required there is no topological difficulty in satisfying the condition (1.12). We may for instance choose the numerical factor for r_j in some arbitrary fashion when $j \leq k/2$ and then define r_j for $j > k/2$ by (1.12). Having defined the $r_j(\xi)$ on the unit sphere S_{n-1} we then define it for all nonzero ξ by making them positive-homogeneous of degree zero.

We now introduce polar coordinates:

$$\xi = \sigma\omega, \qquad \omega \quad \text{on} \quad S_{n-1} .$$

We define $\hat{f}(\sigma, \omega)$ as $\hat{f}(\sigma\omega)$; clearly $\hat{f}$ is an *even* function of $\{\sigma, \omega\}$. However, we define

(1.13) $\phi_j(\sigma, \omega) = \hat{f}(\sigma\omega) \cdot r_j(\omega)$.

Comparing this with (1.8) and using the positive-homogeneity of r_j we see that

$$\phi_j(\sigma, \omega) = \phi_j(\sigma\omega)$$

for σ positive. In general we have

(1.14) $\phi_j(-\sigma, -\omega) = \hat{f}(-\sigma, -\omega) \cdot r_j(-\omega)$

$$= \hat{f}(\sigma, \omega) \cdot r_{j'}(\omega) = \phi_{j'}(\sigma, \omega) ,$$

where again $j' = k - j + 1$.

Since the Fourier transform preserves the L_2 norm we have

$$|f|^2 = \int_{S_{n-1}} \int_0^\infty |\hat{f}(\sigma, \omega)|^2 \sigma^{n-1} \, d\sigma \, d\omega \quad ;$$

and making use of (1.8) and the fact that the $\{r_j\}$ form an orthonormal basis we obtain

$$|f|^2 = \sum_{j=1}^k \int_{S_{n-1}} \int_0^\infty |\phi_j(\sigma, \omega)|^2 \sigma^{n-1} \, d\sigma \, d\omega.$$

Finally, if we employ the relation (1.14) we can rewrite the right side by restricting the summation from 1 to $k/2$ but extending the σ-integration to the whole σ-axis. We get

$$(1.15) \qquad |f|^2 = \sum_{j=1}^{k/2} \int_{S_{n-1}} \int_{-\infty}^\infty |\phi_j(\sigma, \omega)|^2 \sigma^{n-1} \, d\sigma \, d\omega \,.$$

We can restate (1.15) as follows: The correspondence

$$(1.16) \qquad f \leftrightarrow \{\sigma^{(n-1)/2} \phi_j(\sigma, \omega) \; ; j = 1, \cdots, k/2\}$$

is an isometric mapping of H_0 into the $k/2$ fold replica of $L_2(R \times S_{n-1})$. We shall think of the latter function space as

$$L_2(-\infty, \infty; N)$$

where

$$(1.17) \qquad N = [L(S_{n-1})]^{k/2} \,.$$

Since all steps are reversible (at least for the dense set of $\{\phi_j\}$'s which are smooth with compact support excluding $\sigma = 0$), we see that the mapping (1.16) is onto $L_2(-\infty, \infty; N)$ and is thus a unitary map.

Next we calculate the image of $U_0(t)f$ under the mapping (1.16):

$$U_0(t)f \to \{\sigma^{(n-1)/2} \psi_j(\sigma, \omega; t) : j = 1, \cdots, k/2\} \,.$$

Using the explicit expression for $U_0(t)f$ given in (1.9), the definition (1.13) and the relation (1.16) we get

$$(1.18) \qquad \psi_j(\sigma, \omega; t) = \exp(i\sigma\tau_j(\omega)t)\phi_j(\sigma, \omega).$$

Thus, except for the intrusion of the factor $\tau_j(\omega)$ in the exponential, (1.16) behaves like a spectral representation for the group $\{U_0(t)\}$. The passage to a full-fledged spectral representation is easily accomplished, as we show a little later, by a stretching of the variables. But for the present we do not discard (1.16); rather, we put it back into vectorial form by setting

$$(1.19) \qquad \tilde{f}(\sigma, w) = \sigma^{(n-1)/2} \sum_{j=1}^{k/2} \phi_j(\sigma, \omega) r_j(\omega) .$$

For want of a better word we shall call the correspondence

$$f(x) \leftrightarrow \tilde{f}(\sigma, \omega)$$

the pre-spectral representation of H_0 for $\{U_0(t)\}$.

We obtain the pre-translation representer k of f by Fourier transformation:

$$(1.20) \qquad k(s, \omega) = \frac{1}{\sqrt{2\pi}} \int_{-\infty}^{\infty} e^{-is\sigma} \tilde{f}(\sigma, \omega) \, d\sigma .$$

It follows from (1.18) and (1.20) that the pre-translation representer of $U_0(t)f$ is

$$(1.21) \qquad \sum_{j=1}^{k/2} k_j(s - \tau_j(\omega)t, \omega) r_j(\omega) ,$$

where

$$k_j(s, \omega) = k(s, \omega) \cdot r_j(\omega) .$$

This representation

$$f \leftrightarrow \{k_j(s, \omega) ; j = 1, \cdots, k/2\}$$

is also unitary onto $L_2(-\infty, \infty, N)$.

Returning to (1.19), we see by (1.14) that the even [or odd, depending on the parity of $(n-1)/2$] part of $\tilde{f}(\sigma, \omega)$ is just $\sigma^{(n-1)/2}$ times the Fourier transform of f. The latter is the vectorial version of the acoustic spectral representer of $\{0, f\}$ as given by formulas (2.3) and (2.4) of Chapter IV. Consequently the even (or odd) part of the present pre-translation representation is equal to the vectorial version of the translation representation of Chapter IV. We can therefore use Theorem 2.2 of Chapter IV to obtain the following explicit relations between f and its pre-translation representer k:

$$(1.22) \qquad k(s, w) = \sum_{1}^{k/2} (\partial_s^{(n-1)/2} M(s, \omega) \cdot r_j(\omega)) r_j(\omega) ,$$

where

(1.23) $$M(s, \omega) = \frac{1}{(2\pi)^{(n-1)/2}} \int_{x\omega=s} f(x) \, dS$$

and

(1.24) $$f(x) = \int h(x\omega, \omega) \, d\omega$$

where

(1.25) $$h(s, \omega) = \frac{1}{(2\pi)^{(n-1)/2}} (-\partial_s)^{(n-1)/2} k(s, \omega) \; .$$

Note that f depends only on the even part of h and hence on the even [or odd, depending on the parity of $(n - 1)/2$] part of k.

Formulas (1.22), (1.23) hold for all f in S, the space C^∞ of functions all of whose derivatives tend to zero faster than any power of $|x|^{-1}$; the inversion formulas (1.24) and (1.25) hold for the corresponding dense class of h. Since the relation between k and f is unitary, it follows, just as in Chapter IV, that (1.24) and (1.25) is valid in the sense of distributions for all square integrable k.

Using formula (1.21) for the pre-translation representer of $U_0(t)f$ and (1.24), (1.25) we get that

(1.26) $$u(x, t) = \sum_{1}^{k/2} \int h_j(x\omega - \tau_j(\omega)t, \omega) r_j(\omega) \, d\omega \; .$$

From (1.22), (1.23), and (1.26) we can read off property (c) (Huygens' principle) of Theorem 1.1: For suppose that $f(x)$ is zero outside the ball $\{|x| \le R\}$. It then follows from (1.23) that $M(s, \omega)$ is zero for $|s| > R$, and by (1.22) and (1.25) that so are $k(s, \omega)$ and $h(s, \omega)$. Hence for

(1.27) $$|x| < c_{\min} |t| - R$$

the integrand in (1.26) is zero for all ω; this proves that $u(x, t)$ vanishes for all $\{x, t\}$ inside the cones (1.27) and so completes the proof of Theorem 1.1.

We now define D_+ and D_- to consist of those f in H_0 whose pre-translation representers are zero on the negative, respectively positive, s-axis. One can verify immediately by inspection that these are *outgoing* and *incoming subspaces* in the sense of Chapter IV, that is they satisfy the three properties characterizing such subspaces. Furthermore, it follows from the unitary character of the representation that D_+ *is orthogonal to* D_- . Finally, we deduce from the relations (1.25) and (1.26) that if f belongs to D_+ [or D_-], then $u(x, t) = \{U_0(t)f\}(x)$ vanishes in the cone:

$$\{|x| < c_{\min} t\} \text{ [respectively, } \{|x| < -c_{\min} t\}].$$

This last property could have been used to characterize D_+ and D_- as in Chapters IV and V, but for this we need the converse assertion; our proof of this fact is based on the following theorem which is of independent interest.

Theorem 1.2. *Let* $\tau_1, \cdots, \tau_m$ *be m-continuous positive functions on* S_{n-1} *such that for almost all points* ω *no two of the values* $\{\tau_j(\omega)\}$ *are equal. Let* $l_1, \cdots, l_m$ *be square integrable functions on* $R \times S_{n-1}$ *such that*

$$(1.28) \qquad u(x, t) = \int \sum_{j=1}^{m} l_j(x\omega - \tau_j(\omega)t, \omega)\, d\omega$$

is zero inside some cone

$$(1.29) \qquad\qquad \{|x| < ct\} \qquad (c > 0).$$

Then all of the functions $l_j(s, \omega)$ *are zero almost everywhere for* $s < 0$.

Remark. If all of the functions l_j vanish for $s < 0$, then obviously u vanishes inside the cone (1.29) with

$$c = c_{\min} = \min_{j,\omega} |\tau_j(\omega)|.$$

Theorem 1.2 asserts that the converse of this is true.

Let Ω denote the set of all points ω in S_{n-1} at which no two of the values $\{\tau_j(\omega)\}$ are equal. The complement of Ω is by hypothesis of measure zero and since the τ_j are continuous, it is also a closed subset of S_{n-1}. As a consequence the set of continuous functions with support in Ω is dense in $L_2(S_{n-1})$. Our proof of Theorem 1.2 is based on the following lemma.

Lemma 1.1. *Assume that the* $\{\tau_j\}$ *satisfy the hypothesis of Theorem 1.2 and define* Ω *as above. Then, given m continuous functions* $q_1, \cdots, q_m$ *with support in* Ω *and any positive* ϵ, *there exists a single polynomial* p *in* n *variables such that*

$$(1.30) \qquad\qquad \left| q_j(\omega) - p\left(\frac{\omega}{\tau_j(\omega)}\right) \right| < \epsilon$$

simultaneously for all ω *in* S_{n-1} *and all* $j = 1, \cdots, m$.

Proof.   Consider the space S consisting of the union of m unit spheres S_{n-1} with certain identifications: namely, if for some ω, $\tau_j(\omega) = \tau_k(\omega)$, then we identify the points corresponding to ω on the jth and kth spheres. We define n functions $\{f_i\}$ as follows: At the point ω on the jth sphere we set

$$f_i(\omega) = \frac{\omega_i}{\tau_j(\omega)} \qquad (i = 1, \cdots, n),$$

where ω_i denotes the ith component of ω. With the above identifications the functions $\{f_i\}$ are still single-valued and yet they separate the points of S. Therefore, according to the Stone–Weierstrass theorem the polynomials in $f_1, \cdots, f_n$ are dense in the space of all continuous functions on S. It is clear from the definitions of S and Ω that S contains the union of m Ω's (with no identifications); hence the function q defined as equal to q_j on the jth S_{n-1} for each j is single-valued and continuous on S. This proves the simultaneous approximation statement (1.30).

Proof of Theorem 1.2.   By hypothesis the function u defined by (1.28) is zero at all points in the cone (1.29). Let ϕ be an arbitrary $C_0^\infty(0, \infty)$ function; then the integral

$$(1.31) \qquad \int u(x, t)\phi(t)\, dt$$

vanishes for all x in some neighborhood of $x = 0$. Replacing u in (1.31) by (1.28) we see that

$$(1.32) \qquad \iint \sum_{j=1}^{m} l_j(x\omega - \tau_j(\omega)t, \omega)\phi(t)\, d\omega\, dt$$

$$= \iint \sum_{j=1}^{m} l_j(s, \omega)\phi\left(\frac{x\omega - s}{\tau_j(\omega)}\right) d\omega\, ds$$

also vanishes in some neighborhood of $x = 0$. Therefore, the αth partial derivative of (1.32) with respect to x is zero at $x = 0$ for any multi-index α; that is,

$$(1.33) \qquad \iint \partial_s^{|\alpha|}\phi(-s) \sum_{j=1}^{m} \left[\frac{\omega}{\tau_j(\omega)}\right]^\alpha l_j(s, \omega)\, d\omega\, ds = 0,$$

where $|\alpha| = \alpha_1 + \alpha_2 + \cdots + \alpha_n$. We now write

$$a_\alpha(s) = \int \sum_{j=1}^{m} \left[\frac{\omega}{\tau_j(\omega)}\right]^\alpha l_j(s, \omega) \, d\omega ;$$

then $a_\alpha(s)$ is square integrable in s and

$$(1.34) \qquad\qquad \int \partial_s^{|\alpha|} \phi(-s) a_\alpha(s) \, ds = 0$$

for all ϕ in $C_0^\infty(0, \infty)$. It follows from (1.34) that $a_\alpha(s)$ is a polynomial in s of degree less than $|\alpha|$ for $s < 0$ and since it is also square integrable it vanishes for almost all $s < 0$.

The set E_0 of all $s < 0$ for which some $a_\alpha(s)$ does not equal zero or for which some $l_j(s, \omega)$ is not square integrable in ω over S_{n-1} is clearly of measure zero. For fixed s_0 not in E_0 we consider $\{l_j(s_0, \omega); j = 1, \cdots, m\}$ as a vector-valued function in $[L_2(S_{n-1})]^m$. Since the $a_\alpha(s_0)$ vanish for all multi-indices α, it follows that the inner product of $\{l_j(s_0, \omega)\}$ with all polynomial vectors $\{p(\omega/\tau_j(\omega)); j = 1, \cdots, m\}$ vanish. According to Lemma 1.1 this set of polynomial vectors is dense in $[L_2(S_{n-1})]^m$ and therefore the $\{l_j(s_0, \omega)\}$ vanish for almost all ω in S_{n-1}. Since E_0 is of s-measure zero, the functions $l_j(s, \omega)$ are equal to zero almost everywhere in $(-\infty, 0) \times S_{n-1}$. This completes the proof of Theorem 1.2.

The vanishing of u in the cone (1.29) was used only to deduce that the x derivatives of (1.31) are zero at $x = 0$; this already follows from a more meager assumption about u:

Corollary 1.1. *The conclusion of Theorem 1.2 remains true if instead of assuming u to be zero in the cone (1.29) we merely assume that u is zero in some open set containing the positive t-axis.*

Corollary 1.2. *The conclusion of Theorem 1.2 remains true if instead of assuming the functions $\{l_j\}$ to be square integrable we merely assume that they are the Nth derivatives with respect to s of square integrable functions.*

Proof. We first mollify u; that is we form the convolution of u with some C_0^∞ function ψ of t whose support lies on the negative t-axis. Denote the mollified u by $u^{(\psi)}$:

$$(1.35) \qquad u^{(\psi)}(x, t) = u * \psi = \int u(x, t - r)\psi(r) \, dr.$$

Since ψ is by assumption zero on the positive axis and since u vanishes in the cone (1.29), $u^{(\psi)}$ also vanishes there.

Substituting the expression (1.28) for u in (1.35) we get

$$u^{(\psi)}(x, t) = \iint \sum_{j=1}^{m} l_j(x\omega - \tau_j(\omega)(t - r), \omega)\psi(r) \, d\omega \, dr.$$

Interchanging the order of r and ω integration and summation we can write this in the form

$$u^{(\psi)}(x, t) = \sum_{j=1}^{m} \int l^{(\psi)}(x\omega - \tau_j(\omega)t, \omega) \, d\omega$$

where

$$l_j^{(\psi)} = l_j * \psi_j$$

and

$$\psi_j(s, \omega) = [\tau_j(\omega)]^{-1}\psi(s/\tau_j(\omega)).$$

The essential fact here is that $l_j^{(\psi)}$, being the convolution of a C_0^∞ function with one which is the Nth derivative of a square integrable function, is itself square integrable. Therefore we conclude from Theorem 1.2 that

$$(1.36) \qquad l_j^{(\psi)}(s, \omega) = 0 \qquad \text{for} \quad s < 0.$$

Next we choose a sequence $\{\psi\}$ of mollifiers which tend to the δ-function; that is we choose a sequence of functions whose integrals equal one and whose supports tend to the point $t = 0$. It is well known that

$$(1.37) \qquad l_j = \lim l_j^{(\psi)}$$

in the sense of distribution theory. We may in addition choose the mollifiers so that the support of each ψ lies on the negative axis in which case (1.36) is valid; combining this with (1.37) we get that

$$l_j(s, \omega) = 0 \qquad \text{for} \quad s < 0$$

and all j, as asserted in Corollary 1.2.

We are now in a position to prove the following basic result.

Theorem 1.3. *Let f be an element of H_0 such that $u(x, t) = [U_0(t)f](x)$ is zero in the cone* (1.29); *then f belongs to D_+ . An analogous statement holds for D_- .*

Proof. Assume to begin with that the τ_j's are pairwise distinct for almost all ω on S_{n-1} ; and let v be any vector in C_k . Taking the scalar product of formula (1.26) for $u(x, t)$ with v we obtain

$$(1.38) \qquad u(x, t) \cdot v = \int \sum_{j=1}^{k/2} h_j(x\omega - \tau_j(\omega)t, \omega)v_j(\omega) \ d\omega ,$$

where

$$(1.39) \qquad\qquad v_j(\omega) = r_j(\omega) \cdot v .$$

According to formula (1.25), h_j is the s-derivative of order $(n - 1)/2$ of the square integrable function k_j . Since we have assumed that u vanishes in the cone (1.29), we may, using Corollary 1.2 of Theorem 1.2, conclude that

$$(1.40) \qquad\qquad l_j(x , \omega) = h_j(s , \omega)v_j(\omega) = 0$$

for negative s. Since the vector v was arbitrary, we see that h_j itself (and hence k_j) must be zero for negative s; in view of the definition of D_+ this is just what Theorem 1.3 asserts.

Suppose next that some pairs of τ_j's are equal on an ω-set of positive measure. Since the eigenvalues are algebraic functions of ω, any two τ_j's which are equal on a set of positive measure will coincide throughout S_{n-1} . To allow for this possibility let

$$\tau_1'(\omega) \geq \tau_2'(\omega) \geq \cdots \geq \tau_m'(\omega)$$

be a new ordering of the positive τ_j's which omits duplication of identical τ_j's and set

$$(1.41) \qquad\qquad h_i'(s , \omega)r_i'(\omega) = {\sum}' h_j(s , \omega)r_j(\omega) ,$$

where the summation extends over all j's for which $\tau_j(\omega) \equiv \tau_i'(\omega)$. Since only the vector-valued function $h_i'r_i'$ actually enters in the previous argument, $r_i'(\omega)$ can be normalized in an arbitrary fashion. Formula (1.26) can now be rewritten as

$$u(x,\, t) \;=\; \int \sum_{i=1}^{m} h_i'(x\omega - \tau_i'(\omega)t,\, \omega)\, r_i'(\omega)\; d\omega\,.$$

Here the $\tau_i'(\omega)$ are pairwise distinct almost everywhere and the h_i' are s-derivatives of order $(n-1)/2$ of square integrable functions. Hence the previous argument applies and shows that $h_i'(s,\, \omega)$ is zero for all negative s. Since the $r_j(\omega)$ are orthonormal for each ω, this in turn implies that the $h_j(s,\, \omega)$ are zero for all negative s. This completes the proof of Theorem 1.3.

Corollary 1.3. *If u vanishes in any cone* (1.29), *then u vanishes in the cone*: $\{|\,x\,| < c_{\min} t\}$.

Remark. Suppose that the vector v is such that for $j = 1, 2, \cdots, k/2$, the function $v_j(\omega) = r_j(\omega) \cdot v$ is nonzero except possibly on a set of measure zero. Then the vanishing of $h_j(s,\, \omega)$ on the negative axis follows from (1.40) being valid for this particular v, which in turn follows from the vanishing of (1.38) for this particular v. We will call v a *coupled direction* for the system (1.1) if $r_j(\omega) \cdot v \neq 0$ for all j and almost all ω. In the case of coinciding roots we require this condition to hold for all possible choices of the r_j. The above argument then gives

Corollary 1.4. *Let v be a coupled direction for the system* (1.1) *and let u be a solution of* (1.1). *If the component of $u(x, t)$ in the direction of v vanishes in the cone* (1.29), *then u itself vanishes in the cone*: $\{|\,x\,| < c_{\min} t\}$.

Remark. In general one would expect all directions to be coupled. The occurrence of a decoupling direction indicates some kind of degeneracy.

It will be necessary for us to extend the class of initial data to include distributions for which the corresponding pre-translation representers are also distributions. We note the operator relating k to f through formulas (1.24) and (1.25) by $\mathcal{g}$. Since the formulas defining $\mathcal{g}$ are the same as in the acoustic case in Chapter IV, it is to be expected that the ideas and theorems of Section 3 in Chapter IV can be carried over into the present setting.

We take as the domain of the extension of $\mathcal{g}$ the set of all distributions $k(s,\, \omega)$ in s whose values are square integrable C^+-valued functions on

S_{n-1} ;† here $C^{+}(\omega)$ is the space spanned by

$$r_1(\omega) , \cdots , r_{k/2}(\omega) .$$

We shall denote the jth component of k with respect to this orthonormal system by $k_j = k_j(s, \omega)$.

For k in C_0^{∞}, $f = \mathcal{G}k$ is defined by (1.24) and (1.25); we then extend $\mathcal{G}$ by closure in the weak topology. $\mathcal{G}$ so extended has the following properties:

(1.42a) $$\partial_x \mathcal{G} = \mathcal{G}\omega\, \partial_s ;$$

(1.42b) $$A_0 \mathcal{G} = \mathcal{G}a(\omega)\, \partial_s ;$$

(1.42c) $$U_0(t)\, \mathcal{G} = \mathcal{G} S_0(t)$$

where $S_0(t)$ shifts the jth component k_j by $\tau_j(\omega)t$ in the s variable; if

(1.42d) $k = 0$ for $|s| < r$ then $\mathcal{G}k = 0$ for $|x| < r$;

(1.42e)   for all g in C_0^{∞} and all k in the domain of $\mathcal{G}$

$$(g , \mathcal{G}k) = [\mathcal{R}g , k]$$

where $\mathcal{R}$ is defined by (1.22) and (1.23).

We shall mean by the *symmetric part* of k the even part of $k(s, \omega)$ when $(n - 1)/2$ is even, the odd part when $(n - 1)/2$ is odd. The symmetric part of the translation representers treated in Section 3 of Chapter IV correspond to the second components of the data. We are concerned here

---

† More precisely we require that the domain of $\mathcal{G}$ be dual to the space of C^{+}-valued test functions $\varphi(s, \omega)$ which are C^{∞} in s with compact support and majorized in ω by a square integrable function, uniformly in s. The convergence $\varphi_n \to \varphi$ means that there exists a bounded subset K of R_1 such that supp $\varphi_n \subset K \times S_{n-1}$ and for any integer j there exists an F_j in $L_2(S_{n-1})$ such that $\partial_s^j \varphi_n(s, \omega)/F_j(\omega)$ converges uniformly to $\partial_s^j \varphi(s, \omega)/F_j(\omega)$. It is easy to verify that the mappings

$$\varphi(s, \omega) \to \varphi(s - \tau(\omega)t, \omega)$$

and

$$\varphi(s, \omega) \to \varphi(s\tau(\omega), \omega)$$

are isomorphisms. This is required later on by property (1.42c) and the change of scale introduced in the true translation representation. We also use the fact that for distribution data g with compact support, $\mathcal{R}g$ lies in the domain of $\mathcal{G}$; this can be verified by an argument similar to that used in Remark 4.1. of Chapter IV.

with a vector-valued version of such data. Hence Theorem 3.1 of Chapter IV applied to the symmetric part of k gives

Theorem 1.4. *The nullspace of $\mathcal{g}$ consists of distributions k whose spherical harmonics coefficients $k_m(s)$, defined as*

$$k_m(s) \;=\; \int k(s,\,\omega)\, Y_m(\omega)\; d\omega$$

satisfy the condition:

$$k_m(s) \,+\, (-1)^{m+(n-1)/2} k_m(-s)$$

is a polynomial of degree less than $m + (n - 3)/2$.

Similarly, Corollary 3.1 of Chapter IV applied to the symmetric part of k gives

Corollary 1.5. *$\mathcal{g}k$ is zero for $|x| < r$ if and only if*

$$k_m(s) \,+\, (-1)^{m+(n-1)/2} k_m(-s)$$

is a polynomial of degree less than $m + (n - 3)/2$ in the interval $(-r,\,r)$.

Next we turn to Theorem 3.2 of Chapter IV; restricting l to be symmetric in $\{s,\,\omega\}$ but allowing its values to lie in C_k, we obtain

Theorem 1.5. *Let l be a distribution which is symmetric in $\{s,\,\omega\}$, whose values lie in C_k and which is zero for $|s| > r$. Then $\mathcal{g}l$ is zero for $|x| > r$ if and only if l satisfies the orthogonality conditions*

$$(1.43) \qquad\qquad \left[s^\beta Y_m(\omega)\,,\,l \right] = 0$$

for all $\beta \le m + (n - 3)/2$ and all spherical harmonics Y_m.

We extend the class of initial data as follows: Let f be any distribution; we assert that the initial value problem for equation (1.1) has a unique solution with initial data f. This distribution solution is easily constructed as a weak limit of a sequence of solutions whose initial values tend weakly to f. An even quicker way of constructing the solution as well as proving its uniqueness is to note that since $U_0(t)$ maps C_0^∞ into C_0^∞, its adjoint $U_0(-t)$ maps the space of distributions into itself.

As before we denote the solution u with initial value f by Wf. We shall call a distribution solution u *outgoing* [*incoming*] if it is zero in some cone:

$$\{|x| < ct\} [\{|x| < -ct\}] \qquad (c > 0)\;.$$

Data f is called outgoing [incoming] if Wf; finally, we call f *eventually outgoing* [*initially incoming*] if $U_0(t)f$ is outgoing for t large enough positive [respectively, negative].

Starting with (1.22) and (1.23) which explicitly defines the pre-translation representer $\Re f$ for smooth data f with compact support, we extend $\Re$ by continuity to all distributions g with compact support; to do this we made use of the relation (1.42e) which holds in particular for all C^+-valued C_0^∞ functions k. As in Remark 4.1 of Chapter IV, if ϕ is in $C_0^\infty(R_1)$ then

$$\int \Re g \phi(s)\ ds$$

is a bounded function of ω.

At this point it is convenient to introduce the *true translation representation* by stretching the s variable of the components of the pre-translation representation. More precisely, if f is an element of H_0, k its pre-translation representer, and k_j the jth component of k, we define the true translation representer k_0 as

$$k_0 = \{k_{0,j} ; j = 1, \cdots, k/2\} \qquad \text{in} \quad L_2(-\infty, \infty; N)$$

where

(1.44) $$k_{0,j}(s, \omega) = [\tau_j(\omega)]^{1/2} k_j(s\tau_j(\omega), \omega)$$

and

$$N = [L_2(S_{n-1})]^{k/2}$$

It follows from (1.21) that

$$f \leftrightarrow k_0$$

is indeed a translation representation, that is

$$U_0(t)f \leftrightarrow k_0(s - t, \omega).$$

Furthermore, since the pre-translation representation was unitary, so is this one, thanks to the inclusion of the factor $\tau_j^{1/2}$ in (1.44).

We remark that if k is a distribution in s whose values are square integrable in ω, then so is k_0. On the other hand if k is a distribution in both s and ω, then (1.44) does not define $k_{0,j}$ as a distribution unless the functions $\{\tau_j(\omega)\}$ are C^∞. Note that when the roots $\{\tau_j\}$ are distinct, then the $\{\tau_j\}$ are analytic functions.

Following Chapter IV we shall use the true translation representation to construct an outgoing and an incoming solution to the equation $(A_0 - \mu)f = g$.

Theorem 1.6. *Let g be a distribution with compact support and let μ be any complex number. The equation*

$$(1.45) \qquad\qquad (A_0 - \mu)f = g$$

has two uniquely determined distribution solutions, one eventually outgoing and the other initially incoming; we shall denote these as f_+ and f_- .

The proof of uniqueness of these solutions is the same as in Theorem 4.1 of Chapter IV; so is the construction of the true translation representer k_+ and k_- of f_+ and f_-, respectively. As in Corollary 4.1 of Chapter IV we have

Corollary 1.6.

$$(1.46)_o \qquad k_+(s, \omega) = \begin{cases} 0 & \text{for} \quad s < -r/c_{\min}, \\[2mm] e^{-\mu s} n_+(\omega) & \text{for} \quad s > r/c_{\min}\,; \end{cases}$$

and

$$(1.46)_i \qquad k_-(s, \omega) = \begin{cases} e^{-\mu s} n_-(\omega) & \text{for} \quad s < -r/c_{\min}, \\[2mm] 0 & \text{for} \quad s > r/c_{\min}. \end{cases}$$

Here r is such that g is zero for $|x| > r$; $n_\pm$ are bounded functions of ω.

Arguing as in Remark 4.2 of Chapter IV, we see that if f has a true translation representer k_0 and $(A_0 - \mu)f = g$ vanishes for $|x| > r$, then

$$(1.47) \qquad k_0(s, \omega) = \begin{cases} e^{-\mu s} n_1(\omega) & \text{for} \quad s < -r/c_{\min}, \\[2mm] e^{-\mu s} n_2(\omega) & \text{for} \quad s > r/c_{\min}. \end{cases}$$

We now prove a uniqueness theorem for systems.

Theorem 1.7. *Let f be a square integrable function which satisfies*

$$(1.48) \qquad\qquad (A_0 - i\mu)f = 0$$

for $|x| > r$, μ real and different from zero. Then f itself is zero for $|x| > r\, c_{\max}/c_{\min}$.

Proof. Let k_0 be the true translation representer of f. According to (1.47), $k_0(s)$ is periodic in s for $|s| > r/c_{\min}$ and since it is also square integrable it must in fact vanish identically for $|s| > r/c_{\min}$. As a consequence the pre-translation representer k of f vanishes for $|s| > r_0 = r\, c_{\max}/c_{\min}$.

Next let l denote the symmetric part of k:

$$l(s,\omega) = k(s,\omega) \pm k(-s,-\omega),$$

the sign depending on the parity of $(n-1)/2$. According to the rule (1.42b), the pre-translation representer of $A_0{}^j f$ is

$$(1.49) \qquad [a(\omega)]^j \,\partial_s{}^j k .$$

We operate on (1.48) by $A_0{}^j$, j some integer, and obtain

$$(1.50) \qquad A_0^{j+1} f - i\mu A_0{}^j f = 0$$

for $|x| > r$. The pre-translation representer of (1.50) is

$$(1.51) \qquad a^{j+1} \,\partial_s^{j+1} k - i\mu a^j \,\partial_s{}^j k ;$$

here a abbreviates $a(\omega)$. Since the operator $a(\omega)\,\partial_s$ does not alter parity, the symmetric part of (1.51) is

$$(1.51)_{\text{sym}} \qquad a^{j+1} \,\partial_s^{j+1} l - i\mu a^j \,\partial_s{}^j l .$$

Since l vanishes for $|s| > r_0$ we can apply Theorem 1.5. According to the 'only if' part of this theorem the vanishing of (1.50) for $|x| > r$ implies that $(1.51)_{\text{sym}}$ is orthogonal to $s^\beta Y_m$; that is

$$[s^\beta Y_m , a^{j+1} \,\partial_s^{j+1} l] - [s^\beta Y_m , i\mu a^j \,\partial_s{}^j l] = 0$$

for all $\beta \le m + (n-3)/2$. We integrate by parts $(j+1)$ times with respect to s in the first term, j times in the second term, obtaining

$$(1.52) \quad \beta(\beta-1) \cdots (\beta-j) [s^{\beta-j-1} Y_m , a^{j+1} l]$$

$$= i\mu\beta(\beta-1) \cdots (\beta-j+1) [s^{\beta-j} Y_m , a^j l] .$$

Setting $j = \beta$ makes the left side zero and since $\mu \ne 0$ we infer

$$(1.53)_\beta \qquad [Y_m , a^\beta l] = 0 .$$

Setting $j = \beta - 1$ and using $(1.53)_\beta$ we get

$(1.53)_{\beta-1}$ $\qquad\qquad\qquad\qquad [sY_m , a^{\beta-1}l] = 0 .$

Proceeding recursively we deduce that

$(1.53)_\alpha$ $\qquad\qquad\qquad\qquad [s^{\beta-\alpha}Y_m , a^\alpha l] = 0$

for $\alpha = \beta, \beta - 1, \cdots$; for $\alpha = 0$ we get

$$[s^\beta Y_m , l] = 0 .$$

This is condition (1.43) and therefore according to the 'if' part of Theorem 1.5 it follows that $f = \mathcal{g}l$ is zero for $|x| > r_0$; this proves Theorem 1.7.

2. Solutions of Hyperbolic Systems in an Exterior Domain

We now show that our general theory of scattering is applicable to the mixed initial-boundary value problem for hyperbolic systems of the form

$$(2.1) \qquad\qquad u_t = Au = \sum_{i=1}^{n} a^i u_{x^i} + cu$$

in an exterior domain: here the a^i are $k \times k$ Hermitian symmetric matrix-valued functions of class C^1. We assume that, outside of a sufficiently large ball, say for $|x| > \rho$, the a^i do not depend on x and c is zero. Further, we assume that the *sound speed is never zero*; that is we assume that the matrix

$$\sum_{i=1}^{n} a^i(x)\xi^i$$

is invertible for all x and all real $\xi \neq 0$. Finally, we assume that A is formally skew symmetric; in this case the coefficients satisfy the additional condition

$$(2.2) \qquad\qquad \sum_{i=1}^{n} \partial_{x^i} a^i = c + c^* .$$

The domain G lies in the exterior of its boundary ∂G which is smooth and contained in the ball $\{|x| < \rho\}$.

The mixed problem consists of finding a solution u of (2.1) defined for all x in G and all t whose initial value is prescribed and which satisfies a finite number of linear homogeneous conditions on the boundary. We shall put these boundary conditions in the following form: At each point z of the boundary a subspace $\mathcal{B}(z)$ of C_k is prescribed and we require for all time t that

(2.3) $u(z, t)$ *lies in* $\mathcal{B}(z)$ *at each point z of* ∂G .

We further require that the boundary conditions be such that the solution is *energy preserving*; that is, that the L_2 norm of u over G should be independent of t. For an equation of the form (2.1) this is the case if and only if the operator A is skew symmetric; this means that for all pairs of functions f and g in the domain of A

$$(Af, g) + (f, Ag) = 0 .$$

Using the definition of A we can write the left side of this expression as

$$\int_G \{\sum (a^i f_{x^i} \cdot g + f \cdot a^i g_{x^i}) + cf \cdot g + f \cdot cg\} \, dx.$$

Because of the symmetry of the a^i and the condition (2.2), the integrand is a perfect x derivative and carrying out the integration, we obtain the surface integral

(2.4) $$\int_{\partial G} f \cdot \{\sum a^i n_i\} g \, dS,$$

where the $\{n_i\}$ are components of the outer normal to ∂G. We shall use the abbreviation

$$b = \sum a^i n_i;$$

Hence the solutions of (2.1) will be energy preserving if and only if (2.4) is zero for all f and g in the domain of A, that is for all f and g satisfying the boundary conditions (2.3). This is equivalent to

(2.5) $f \cdot bg = 0$ for all f, g in $\mathcal{B}(z)$ at each point z of ∂G ;

that is, *the quadratic form associated with b is zero on* $\mathcal{B}$.

In order to have the largest possible class of solutions we must require in addition that the boundary conditions be *minimal* in the sense that if we omit a single boundary condition then (2.5) is violated. This can be expressed by the condition that the subspace $\mathcal{B}$ be *maximal* with respect to property (2.5).

As noted in Section 1 the signature of the matrix b is zero; in fact b has $k/2$ positive and $k/2$ negative eigenvalues. The following lemma gives a complete description of all subspaces which have the property (2.5) maximally.

Lemma 2.1.   *Let b be a symmetric $k \times k$ nonsingular matrix of signature zero, and denote by P and N the $k/2$ dimensional subspaces spanned by the eigenvectors corresponding to the positive, respectively negative, eigenvalues. Then b is positive, respectively negative, definite over P and N. Let $p_1 , \cdots , p_{k/2}$ be any orthonormal basis in P with respect to b and $n_1 , \cdots , n_{k/2}$ be any orthonormal basis in N with respect to $-b$. Then*

$$(2.5)' \qquad\qquad f \cdot bf = 0$$

for all f in the subspace $\mathfrak{B}$ spanned by

$$p_1 + n_1 , \cdots , p_{k/2} + n_{k/2}$$

and $\mathfrak{B}$ is maximal with respect to this property. Conversely, every $\mathfrak{B}$ which has property $(2.5)'$ maximally can be constructed in this way.

The proof of Lemma 2.1 is easy and is left to the reader.

Suppose next that at each boundary point z there is prescribed a subspace $\mathfrak{B}(z)$ which has property (2.5) maximally and that $\mathfrak{B}(z)$ depends differentiably on z. We wish to define the operator A subject to the boundary conditions (2.3) so that it will be skew self-adjoint. There are two ways of doing this. We denote by H the Hilbert space of square integrable C_k-valued functions in G with norm

$$|f| = \left\{ \int_G f \cdot f \, dx \right\}^{1/2}.$$

(a).   *The Strong Definition.* A is the closure in H of the operator (2.1) as defined for differentiable functions in H which satisfy the boundary condition (2.3) at every point of the boundary.

(b).   *The Weak Definition.* The domain of A consists of those f in H for which there exists a g in H such that

$$(f , Ah) = -(g , h)$$

for all differentiable h in H which satisfy the boundary conditions (2.3).

Generalizing an earlier result of Friedrichs [1], the authors have shown in [1] that these two definitions of A are equivalent and hence that A so defined is skew self-adjoint. Thus, according to Stone's theorem the operator A generates a group of unitary operators

$$U(t) = \exp At$$

relating the initial values of solutions of (2.1) to their values at time t.

In order to study the scattering operator for the group $\{U(t)\}$ with respect to the unperturbed group $\{U_0(t)\}$ we also require that the boundary conditions (2.3) be *coercive* in the sense of Aronszajn:

Definition 2.1. A boundary condition is called coercive for an operator A if there exists two positive constants C, C' such that

(2.6) $$| \partial_x f | \leq C | Af | + C' | f |$$

for all functions f defined on G which satisfy the boundary condition. Here $| f |$ denotes the L_2 norm of f over G.

The following is an example of a noncoercive boundary condition: $k = 2$ so that $f = \{ f_1 , f_2 \}$ and G is a domain in R_2. The boundary condition is $f_1 = if_2$ and

$$A = \begin{pmatrix} 0 & 1 \\ 1 & 0 \end{pmatrix} \partial_x + \begin{pmatrix} 1 & 0 \\ 0 & -1 \end{pmatrix} \partial_y .$$

A brief calculation shows that this boundary condition is minimal and energy preserving. On the other hand, as is easily verified, if h is any analytic function of $z = x + iy$, then $f = \{ih, h\}$ is annihilated by A. For such an f the coerciveness inequality (2.6) would assert that

$$| \partial_z h | \leq \text{const} | h |$$

which is obviously not valid for all analytic functions on G.

There is a rather simple algebraic criterion for deciding whether or not a boundary condition is coercive. As a first step in establishing this we perform a few analytic simplifications.

(a). By employing a partition of unity we can reduce the problem to that of deciding whether the coerciveness inequality (2.6) holds for functions f with small support.

(b).   Since a small portion of the boundary of G can be mapped into a hyperplane, it suffices to consider the case where the domain is a half-space with points $\{z, y\}$, z in R_{n-1} and $y > 0$.

(c).   Let x_0 be any point in the support of f and let A_0 denote the operator

$$A_0 = \sum_{i=1}^{n} a^i(x_0)\, \partial_{z^i}\,.$$

Clearly,

(2.7) $| Af - A_0 f | < \epsilon\, | \partial_x f | + \text{const}\, | f |\,,$

where ϵ is the *oscillation* of the coefficients $a^i(x)$ on the support of f; ϵ is small when the support of f is small.

The coerciveness inequality for the operator A_0 should be of the form

(2.6)$_0$ $| \partial_x f | \leq C_0\, | A_0 f | + C_0'\, | f |\,.$

For ϵC and ϵC_0 both less than one it is obvious that (2.6)$_0$ and (2.7) imply (2.6), and also that (2.6) and (2.7) imply (2.6)$_0$; consequently (2.6) and (2.6)$_0$ are equivalent.

The problem is therefore reduced to finding conditions under which (2.6)$_0$ holds in a half-space. We first write A_0 in terms of the tangential and normal coordinates $\{z, y\}$:

(2.8) $A_0 = \sum_{i=1}^{n-1} c^i\, \partial_{z^i} + b\, \partial_y\,.$

Since this is a homogeneous operator with constant coefficients, its null space contains exponential functions:

$$g(z\,,\,y)\ =\ \exp(z\zeta + y\eta)v\ =\ \exp(l(z\,,\,y))v\,,$$

where ζ lies in C_{n-1} , η in C, v in C_k , and $l = z\zeta + y\eta$. Then

$$A_0 g\ =\ e^l[c(\zeta) + b\,\eta]v\,,$$

where we have set

$$c(\zeta)\ =\ \sum_{i=1}^{n-1} c^i\zeta^i\,.$$

The condition $A_0 g = 0$ means therefore that

$$(2.9) \qquad\qquad [c(\zeta) + b\eta]v = 0 .$$

This relation is satisfied by a nonzero v if and only if

$$(2.10) \qquad\qquad \det [c(\zeta) + b\eta] = 0 .$$

We take ζ to be purely imaginary and different from zero. Then η cannot be purely imaginary since this would contradict the basic assumption that for real nonzero ξ the matrix $\sum a^i \xi^i$ is nonsingular. Therefore the k roots of the algebraic equation (2.10) split into two classes: those with negative real parts and those with positive real parts. The roots as well as the number of roots in each class varies continuously with ζ. Assuming $n - 1$ to be greater than one we can deform ζ into $-\zeta$ without going through the origin; and since the roots are odd functions of ζ, it follows that exactly half of the roots lie in the left half-plane; we denote these by

$$\eta_1(\zeta) , \cdots , \eta_{k/2}(\zeta) .$$

We denote the corresponding null vectors by

$$v_1(\zeta) , \cdots , v_{k/2}(\zeta) .$$

Finally we denote the space spanned by $v_1, \cdots , v_{k/2}$ as $\mathcal{E}(\zeta)$. The roots lying in the right half-plane and their corresponding null vectors are similarly denoted with subscripts running from $k/2 + 1$ through k.

Lemma 2.2.   *The boundary condition* (2.3) *is coercive if and only if for any nonzero purely imaginary ζ the boundary space $\mathcal{B}$ and the space $\mathcal{E}(\zeta)$ have only the zero vector in common.*

Remark.   Since $b^{-1}c(\zeta)$ need not be (skew) symmetric some of the null vectors above may be generalized null vectors. In this case the proof of Lemma 2.2 has to be modified somewhat. However, with this interpretation of the v_j's and $\mathcal{E}$, the assertion remains valid.

Proof.   We begin by proving the 'only if' part of the lemma. Suppose that for some ζ there is a v in $\mathcal{B} \cap \mathcal{E}(\zeta)$; then v is of the form

$$v = \sum_{j=1}^{k/2} \beta_j v_j$$

where the β_j are complex numbers. The function

$$(2.11) \quad g(z, y) = \sum \beta_j \exp(l_j) v_j = \exp(z\zeta) \sum \beta_j \exp(y\eta_j) v_j,$$

where $l_j = z\zeta + y\eta_j$, is equal to $\exp(z\zeta)v$ on the boundary set: $\{y = 0\}$ and hence g satisfies the boundary condition (2.3). Also, g satisfies

$$A_0 g = 0$$

since g is the sum of functions each of which is annihilated by A_0.

We now choose any scalar function $\phi = \phi(z, y)$ which is differentiable, has compact support and is equal to one in the neighborhood of some boundary point. Let α denote any positive number; we define f_α by

$$f_\alpha = \phi \sum_{j=1}^{k/2} \exp(\alpha l_j) \beta_j v_j \equiv \phi g_\alpha.$$

Since the relation (2.9) is homogeneous it follows that g_α is annihilated by A_0 and using the Leibniz rule, we get

$$(2.12) \qquad A_0 f_\alpha = A_0[\phi g_\alpha] = \phi A_0 g_\alpha + (A_0\phi) g_\alpha = (A_0\phi) g_\alpha.$$

The function ϕ has compact support and g_α is exponentially decreasing:

$$|g_\alpha(z, y)| \le \text{const } e^{-\alpha\mu y}$$

in absolute value, where $-\mu$ is the *maximum* of the real parts of $\{\eta_j ; j = 1, \cdots, k/2\}$. Using these facts we easily deduce that

$$(2.13) \qquad\qquad\qquad |f_\alpha| = |\phi g_\alpha| \le \frac{\text{const}}{\alpha^{1/2}}$$

in norm and, using (2.12), that

$$(2.14) \qquad\qquad\qquad |A_0 f_\alpha| = |(A_0\phi) g_\alpha| \le \frac{\text{const}}{\alpha^{1/2}}$$

also in norm. On the other hand $\phi \equiv 1$ in some neighborhood of a boundary point and since the v_j are linearly independent we have

$$|\partial_z g_\alpha| = |\alpha\zeta g_\alpha(z, y)| \ge \text{const } \alpha e^{-\alpha\nu y}$$

in absolute value, where $-\nu$ is the *minimum* of the real parts of $\{\eta_j\}$. Combining these facts we conclude that

$$(2.15) \qquad |\partial_x f| \geq |\partial_z f| \geq \text{const } \alpha^{1/2}$$

in norm. The inequalities (2.13), (2,14), and (2.15) show that no inequality of the form $(2.6)_0$ is satisfied by f_α for all α. This proves the 'only if' part of Lemma 2.2.

To prove the 'if' part we have to deduce the inequality $(2.6)_0$; here we merely sketch a proof. Setting $h = A_0 f$, we take the Fourier transform with respect to the z variables; this gives

$$(2.16) \qquad -ic(\xi)\hat{f} + b\,\partial_y\hat{f} = \hat{h}\,,$$

where $\hat{f}$ denotes the z-Fourier transform of f and ξ, dual to z, is real-valued. The relation (2.16) is an ordinary differential equation in y which is easily solved by the method of normal modes: take the inner product of (2.16) with v_j. Making use of the symmetry of $c(\xi)$ and b this can be written as

$$-i\hat{f}\cdot c(\xi)v_j + \partial_y\hat{f}\cdot bv_j = \hat{h}\cdot v_j\,.$$

From equation (2.9) (with $i\xi$ in place of ζ) we see that we can rewrite the above equation in the form

$$(2.17) \qquad -\bar{\eta}_j\hat{f}_j + \partial_y\hat{f}_j = \hat{h}_j\,;$$

here we have used the abbreviations

$$\hat{f}_j = \hat{f}\cdot bv_j \qquad \text{and} \qquad \hat{h}_j = \hat{h}\cdot v_j\,.$$

The functions $\{\hat{h}_j\}$ have compact support in the y variable.

For Re η_j positive, equation (2.17) has exactly one solution which is square integrable on the positive real axis; whereas for Re η_j negative all solutions of (2.17) are square integrable on the positive real axis. Hence the differential equation (2.16) always has an L_2 solution if h is square integrable. All other L_2 solutions can be obtained by adding L_2 solutions of the homogeneous equation; these are just linear combinations of

$$\{\exp(\eta_j y)v_j\,;j = 1\,,\cdots,k/2\}\,.$$

The value at $y = 0$ of such a linear combination may be any element of $\mathcal{E}(\zeta)$, but $\hat{f}(0)$ must lie in $\mathcal{B}$. If the hypothesis of Lemma 2.2 is fulfilled, that is if $\mathcal{E}(\zeta) \cap \mathcal{B} = \{0\}$, then since both $\mathcal{E}$ and $\mathcal{B}$ are $k/2$-dimensional it follows that $\mathcal{E} + \mathcal{B} = C_k$. This shows that the differential equation (2.16) has a unique solution whose value at $y = 0$ belongs to $\mathcal{B}$.

Making these qualitative arguments quantitative, we can derive not only the existence of $\hat{f}$ but also the following estimate for $\hat{f}$:

$$\int \{\xi^2 |\hat{f}|^2 + |\partial_y \hat{f}|^2\} \, dy \leq \text{const} \int |\hat{h}|^2 \, dy + \text{const} \int |\hat{f}|^2 \, dy \, .$$

Integrating this with respect to ξ and using the L_2 norm-preserving character of the Fourier transform, we obtain $(2.6)_0$. This completes our discussion of Lemma 2.2.

Our problem has now been reduced to finding boundary conditions which are minimal, energy preserving, and coercive or, equivalently, to finding subspaces $\mathcal{B}$ of dimension $k/2$ such that

(i) $u \cdot bu = 0$ for all u in $\mathcal{B}$,
(ii) $\mathcal{B} \cap \mathcal{E}(\zeta) = \{0\}$ for all purely imaginary $\zeta \neq 0$.

Are these two conditions compatible? To raise some doubt in the reader's mind we point out that the spaces $\mathcal{E}(\zeta)$, which are spectacularly non-coercive, are themselves energy preserving. To see this we take the equation (2.9):

$$c(\zeta)v_j + \eta_j b v_j = 0$$

and form the inner product of it with v_l:

(2.18) $c v_j \cdot v_l + \eta_j b v_j \cdot v_l = 0 \, .$

Interchanging the indices j and l we get

(2.18)' $c v_l \cdot v_j + \eta_l b v_l \cdot v_j = 0 \, .$

Since ζ is purely imaginary the matrix $c = c(\zeta)$ is skew symmetric; b is symmetric. Therefore, if we take the complex conjugate of (2.18)' and add it to (2.18), then the first two terms add up to zero and we obtain

(2.19) $(\eta_j + \bar{\eta}_l) b v_j \cdot v_l = 0 \, .$

Since both η_j and $\bar{\eta}_l$ have negative real parts, so does their sum; that is $\eta_j + \bar{\eta}_l \neq 0$. We deduce from (2.19) that

$$b v_j \cdot v_l = 0 \, .$$

It follows from this that for any linear combination u of the $\{v_j ; j = 1, \cdots, k/2\}$, $u \cdot bu = 0$. This shows that $\mathcal{E}(\zeta)$ is indeed energy preserving.

The following is an example of an operator for which *no energy preserving boundary condition is coercive*: $k = 2$, G is the half-plane $\{x_3 > 0\}$ in R_3, and the operator A_0 is

$$(2.20) \qquad\qquad A_0 = \sum_{j=1}^{3} \sigma_j u_{x^j} ,$$

where

$$(2.20)' \quad \sigma_1 = \begin{pmatrix} 0 & 1 \\ 1 & 0 \end{pmatrix}, \quad \sigma_2 = \begin{pmatrix} 0 & -i \\ i & 0 \end{pmatrix}, \quad \sigma_3 = \begin{pmatrix} 1 & 0 \\ 0 & -1 \end{pmatrix}.$$

Taking $c^1 = \sigma_1$, $c^2 = \sigma_2$, and $b = \sigma_3$, two brief calculations show that

(a)   Every energy conserving space $\mathcal{B}$ consists of complex multiples of a vector of the form $\{1, \theta\}$ for fixed θ with $|\theta| = 1$.

(b)   For $\zeta = i\omega$ with $|\omega| = 1$, we get $\eta_1 = -1$ and $v_1 = \{1, \omega_2 - i\omega_1\}$. Clearly, for $\omega_2 - i\omega_1 = \theta$ the space $\mathcal{E}(\zeta)$ coincides with $\mathcal{B}$.

In what follows we assume that we are dealing with boundary conditions satisfying (i) and (ii) in which case the inequality (2.6) holds. Combining this with the Rellich compactness criterion we deduce:

Lemma 2.3.   *Let F denote the set of all f such that*

$$|Af| + |f| \leq 1 ;$$

this set F is precompact with respect to the L_2 norm over any bounded subdomain of G.

A property of solutions of (2.1) which plays a crucial role in our arguments in the finiteness of the speed with which signals propagate:

Lemma 2.4.   *If u is a solution of (2.1) and $u(x, 0) = 0$ for $\{|x| > r\}$, then $u(x, t) = 0$ for $\{|x| > c_{\max}|t| + r\}$.*

We come now to the notion of incoming and outgoing subspaces and here we have two different definitions both of which seem quite natural. As in Chapter V we call a solution $u(x, t)$ outgoing [incoming] if for $t > 0$

$[t < 0]$ the signal u is zero in the scattering region $\{|x| < \rho\}$; the initial values of outgoing and incoming solutions constitute the outgoing and incoming subspaces $D_+(\rho)$ and $D_-(\rho)$, respectively.

We remark that outgoing and incoming solutions may be regarded as free space solutions for $t > 0$, $t < 0$, respectively. Therefore, it follows from part (b) of Theorem 1.1 and Corollary 1.3 of Theorem 1.3 that such solutions are zero in the cone:

$$\left\{|x| < \rho\,\frac{c_{\min}}{c_{\max}} \pm c_{\min}t\right\}.$$

From the point of view of relating the incoming and outgoing representation of the perturbed problem to those for the free space problem, however, it is convenient to define the incoming and outgoing subspaces of the former in reference to those of the latter as follows:

$$D_+{}^\rho = U_0(\rho/c_{\min})D_+,$$

$$D_-{}^\rho = U_0(-\rho/c_{\min})D_-.$$

It is clear from these definitions that

$$D_+{}^\rho \subset D_+(\rho) \subset D_+ \qquad \text{and} \qquad D_-{}^\rho \subset D_-(\rho) \subset D_-.$$

Hence in the sense of Definition 6.1 of Chapter III, $D_+{}^\rho$ and $D_+(\rho)$ are equivalent outgoing subspaces and $D_-{}^\rho$ and $D_-(\rho)$ are equivalent incoming subspaces. It follows therefore from Theorem 6.2 of Chapter III that the scattering matrices associated with the two pairs are equivalent; that is they differ from each other by at most entire and invertible factors.

The subspaces $D_\pm{}^\rho$ obviously have properties (i) and (ii) characterizing incoming and outgoing subspaces. Moreover, being subspaces of the orthogonal pair D_+ and D_-, they themselves are orthogonal.

Property (iii)$_+$, namely,

(2.21) $\qquad\qquad \cup\, U(t)D_+{}^\rho \qquad$ is dense in $\quad H,$

lies deeper, however. Paralleling the approach in Chapter V, we prove (iii) in two steps:

(1). If A has no point spectrum, then using the local compactness property expressed in Lemma 2.3 we can deduce that energy decays locally; that is, that

$$(2.22) \qquad\qquad \liminf_{t\to\infty} \int_{G'} |u(x, t)|^2\, dx = 0$$

for all bounded subdomains G' of G.

(2)   Making use of the finiteness of sound speed and Huygens' principle in free space we can deduce (2.21) from (2.22).

The only gap in this scheme concerns the point spectrum of A; we need to show that

$$(2.23) \qquad\qquad\qquad Af = i\mu f$$

has no square integrable nontrivial solution for real μ. We deal with the case $\mu \neq 0$ first:

For $|x| > \rho$ we have assumed that the coefficients of A are constant and the zero order term vanishes. Thus we can write (2.23) in the form:

$$(2.23)_0 \qquad\qquad (A_0 - i\mu)f = 0 \qquad \text{for} \quad |x| > \rho\,;$$

here A_0 is an operator of the kind treated in Section 1. According to Theorem 1.7 any square integrable solution of $(2.23)_0$ is necessarily zero for all $|x| > \rho\, c_{\max}/c_{\min}$. From this and Eq. (2.23) we can conclude that f is zero for all x in G provided that Eq. (2.23) has the *unique continuation property*, which means that a solution of (2.23) zero on an open subset is zero everywhere in G. Now the following class of operators are known to have the unique continuation property:

(a)   Operators with analytic coefficients,
(b)   Operators with simple complex characteristics.

From now on we shall consider only operators A which fall into one of these two classes. For such operators $i\mu$ with $\mu \neq 0$ is not an eigenvalue of A. The possibility of $\mu = 0$ being an eigenvalue remains. Unlike the acoustic case we can not exclude this possibility and therefore if we insist on property (iii) we have to restrict the discussion to the orthogonal complement H' of the null space of A.

Lemma 2.5. D_+^ρ and D_-^ρ belong to H'.

Proof.   We must show that for every g in say D_+^ρ and every f in the nullspace of A

(2.24) $$(g\,,f)\ =\ 0\ .$$

Since $Af\ =\ 0$, it follows that

$$\frac{d}{dt}\,U(t)f\ =\ U(t)Af\ =\ 0$$

so that

(2.25) $$U(t)f\ =\ f$$

for all t. Using the isometry of $U(t)$ and (2.25) we have

(2.26) $$(g\,,f)\ =\ (U(t)g\,,\,U(t)f)\ =\ (U(t)g\,,f)\ .$$

Since g was assumed to lie in $D_+{}^\rho$, $U(t)g$ vanishes for $|\,x\,|\ <\ \rho\ +\ c_{\min}t$. Applying the Schwarz inequality we therefore get

$$|\,(U(t)g,f)\,|\ \leq\ |\,U(t)g\,|\,\int\,|\,f(x)\,|^2\,dx\,,$$

where the domain of integration extends over the set $\{|\,x\,|\ \geq\ \rho\ +\ c_{\min}t\}$. The first factor $|\,U(t)g\,|$ on the right is independent of t and the second factor tends to zero as t becomes infinite. Combining this with (2.26) we conclude that

(2.27) $$(g\,,f)\ =\ \lim_{t\to\infty}\ (U(t)g\,,f)\ =\ 0\ ;$$

this proves Lemma 2.5.

We summarize what we have proved so far:

Theorem 2.1. *If the boundary condition* (2.3) *imposed on A is minimal, energy preserving, and coercive, and if A belongs to one of the two classes of operators which have the unique continuation property, then $D_-{}^\rho$ and $D_+{}^\rho$ are orthogonal incoming and outgoing subspaces of H'; here H' is the orthogonal complement of the nullspace of A in H.*

It follows then that H' has incoming and outgoing translation and spectral representations. The latter are related by the scattering matrix, which is bounded and analytic in the lower half-plane.

As pointed out in Chapter V, property (iii) implies the strong form of the local energy decay:

Corollary 2.1.   *Let f be in H'. Then given any bounded subset G' of G and any positive ϵ, there exists T such that*

$$\int_{G'} |\, U(t)f\,|^2\, dx < \epsilon$$

for all t > T.

In order to obtain further properties of the scattering matrix we now study the semigroup

$$Z^\rho(t) \;=\; P_+{}^\rho U'(t) P_-{}^\rho$$

where, as before, $P_\pm{}^\rho$ are the orthogonal projections onto the complements of $D_\pm{}^\rho$ and $U'(t)$ denotes the restriction of $U(t)$ to H'. The domain of $Z^\rho(t)$ is $K^\rho = H' \ominus (D_+{}^\rho \oplus D_-{}^\rho)$ and we denote the infinitesimal generator of $\{Z^\rho(t)\}$ by B^ρ.

In analogy to Theorem 3.1 of Chapter V we now have

Theorem 2.2.   *The operator*

$$Z^\rho(2\rho/c_{\min})\,(\kappa I - B^\rho)^{-1}$$

is compact for any $\kappa > 0$.

The proof, based on elementary properties of the free space translation representation, the finiteness of sound speed and local compactness as expressed by Lemma 2.3, is exactly the same as that of its counterpart in Chapter V.

According to the abstract theory presented in Chapter III (Corollaries 4.2 and 4.5), it follows from Theorem 2.2 of this chapter that *the generator B^ρ has a pure point spectrum and that the resolvent of B^ρ is meromorphic in the plane, holomorphic in an open set including the right half-plane and the imaginary axis.*

Since the semigroup $\{Z^\rho(t)\}$ was obtained from the group $\{U'(t)\}$ by multiplication on the left and right by certain projection operators, the infinitesimal generators B^ρ and A will be, roughly speaking, similarly related. This suggests that the resolvent set of B^ρ might also be a resolvent set for A. If we extend A to act on all f which are merely locally in the domain of A, this is indeed so in the following precise sense:

Theorem 2.3. *Suppose that μ belongs to the resolvent set of B^ρ. Let g be any function in H' which vanishes for $|x| > c \geq \rho$ (or, more generally, let g belong to K^c). Then the equation*

$$(2.28) \qquad\qquad (A - \mu)f = g$$

has a uniquely determined solution

$$f = f_+(\mu)$$

which is eventually outgoing. Similarly, there is a uniquely determined initially incoming solution.

We recall from section one that eventually outgoing means that $U_0(t)f$ vanishes in some forward cone: $\{|x| < c_{\min}(t - b)\}$.

Theorem 2.3 is the main existence theorem of this subject. A proof can be read off from that of Theorem 4.2 of Chapter V; the construction of f which we sketch below is somewhat more direct.

Since the operator A is skew self-adjoint, $A - \mu$ is invertible as an operator on H' for all μ not on the imaginary axis. When $\operatorname{Re} \mu > 0$ we claim that

$$(2.29)_+ \qquad\qquad f_+ = (A - \mu)^{-1}g$$

is eventually outgoing. To see this we recall that

$$f_+ = (A - \mu)^{-1}g = \int_0^\infty e^{-\mu t}U'(t)g\,dt \qquad \text{for} \quad \operatorname{Re}\mu > 0\,.$$

Since D_-^c is mapped into itself by $U(-t)$ if $t > 0$, it follows that the orthogonal complement is mapped into itself by $U(t)$; thus $U(t)g$ and hence f_+ is orthogonal to D_-^c. Making use of the free space translation representation and the definition of D_-^c, we see that $U_0(c/c_{\min})f_+$ lies in D_+ and therefore f_+ is eventually outgoing.

Similarly, we can show that $(A - \mu)^{-1}g$ is initially incoming for $\operatorname{Re} \mu < 0$ and therefore can not be used for $f_+(\mu)$. It turns out that $f_+(\mu)$ can be defined as the local analytic continuation of $\{(A - \mu)^{-1}g; \operatorname{Re}\mu > 0\}$; we carry this out in the weak sense as follows:

In the definition of $\{Z^\rho(t)\}$ we can replace ρ by any $a > \rho$. According to Theorem 6.3 of Chapter III the generator B^a of the resulting semigroup has the same spectrum as B^ρ. With this in mind let a be any number $> c$ and let h be any L_2 function with support in the ball $\{| \, x \, | < a \, c_{\min}\}$. Then h can be decomposed into orthogonal parts:

$$h = h' + h_0 \, ,$$

where h' lies in H' and h_0 in the nullspace of A. Now h itself is obviously orthogonal to $D_+{}^a$ and by Lemma 2.5 so is h_0 ; thus,

$$P_+{}^a h' = h' \, .$$

Since g belongs to K^c which is contained in K^a we also have

$$P_-{}^a g = g \, .$$

Hence, the formula

$$(f_+(\mu) \, , h) = ((A - \mu)^{-1} g \, , h)$$

valid for $\operatorname{Re} \mu > 0$ can be rewritten as follows:

$$(f_+(\mu) \, , h) = (f_+(\mu) \, , h') = ((A - \mu)^{-1} P_-{}^a g \, , P_+{}^a h')$$

$$= (P_+{}^a (A - \mu)^{-1} P_-{}^a g \, , h') = ((B^a - \mu)^{-1} g \, , h')$$

(2.30) $\qquad = ((B^a - \mu)^{-1} g \, , h) \, ;$

here we have used the fact that h_0 is orthogonal to $f_+(\mu)$ and $(B^a - \mu)^{-1} g$. Clearly the quantity on the extreme right can be continued analytically to the resolvent set of B^ρ; therefore, so can the quantity on the extreme left. This serves to continue $f_+(\mu)$ analytically for all μ in the resolvent set of B^ρ as a locally square integrable function.

It remains to be shown that $f_+(\mu)$, thus continued analytically, satisfies the differential equation (2.28) and is eventually outgoing. To accomplish this we must express these conditions in a weak sense. Fortunately this is easily done; using the weak definition of the operator A we know that

(1) $f_+(\mu)$ satisfies (2.28) if and only if

(2.31) $\qquad\qquad (f_+(\mu) \, , (A - \bar{\mu})h) = (g \, , h)$

for all C^∞ data h with bounded support which satisfies the boundary condition adjoint to (2.3).

(2) $f_+(\mu)$ is eventually outgoing if and only if

(2.32) $\qquad\qquad (f_+(\mu) \, , U_0(-t)h) = 0$

for every square integrable function h which is zero for $\{|x| < c_{\min}(t - b)\}$, b fixed.

We know that (2.31) and (2.32) are true for $\mathrm{Re}\,\mu > 0$; they clearly remain true under analytic continuation by (2.30). This completes our construction of f_+.

Knowing the existence of a solution $f_+(\mu)$ of (2.28) we shall prove its uniqueness as well as give a more effective method for its construction in the following:

Theorem 2.4. *Suppose that μ does not belong to the spectrum of B^ρ and let g be any square integrable function with bounded support in G. Denote by $u(x, t)$ any solution of the inhomogeneous equation*

$$(2.33) \qquad u_t - Au = e^{\mu t}g$$

whose initial values are square integrable. Then as t tends to infinity $ue^{-\mu t}$ tends to $f_+(\mu) + w$ in the local energy norm where w is some square integrable solution of $Aw = 0$.

Remark. This result is called the *principle of limiting amplitude*.

The proof is based on the following property of f_+ which we prove below: f_+ can be decomposed as a sum

$$(2.34) \qquad f_+ = f_1 + f_2$$

where f_1 is in L_2 and f_2 is in $D_+(b)$.

Proof. The function f_+ satisfies Eq. (2.28); it follows from this and the coercive inequality (2.6) that f_+ and its first partial derivatives are locally square integrable. Therefore, f_+ also satisfies an equation of the form

$$(2.35) \qquad (A_0 - \mu)f_+ = g_0,$$

where g_0 is square integrable and vanishes for $|x| > c$.

In addition, f_+ is eventually outgoing, and according to Theorem 1.6 the eventually outgoing solution of (2.35) has a free space pre-translation representer k_+ which is zero for $s < -c/c_{\min}$ and, going back to formula (4.6)$'$ of Chapter IV, we see that it is locally square integrable. The corresponding pre-translation representer k of f_+ is locally square integrable and vanishes for $s < -b$ where $b = c\,c_{\max}/c_{\min}$. Decompose k as

$$k = k_1 + k_2,$$

where k_1 is taken to be zero for $s > b$ and k_2 to be zero for $s < b$. Clearly, k_1 is square integrable and therefore represents a square integrable function f_1 while, by properties (1.42c) and (1.42d), $f_2 = gk_2$ is such that $U_0(t)f_2 = 0$ for $|x| > b + c_{\min}t$; we abbreviate this by saying that f_2 is in $D_+(b)$.

We now return to the function u defined as the solution of the inhomogeneous equation (2.33). The function $e^{\mu t}f_+(\mu)$ is also a solution; therefore the difference

$$v = e^{\mu t}f_+(\mu) - u$$

satisfies the homogeneous equation

$$v_t - Av = 0 .$$

Thus we can write

(2.36) $$v(t) = U(t)v(0) .$$

By the definition of v

$$v(0) = f_+ - u(0) .$$

Since $u(0)$ is in L_2 by assumption, the decomposition (2.34) of f_+ shows that $v(0)$ can be written as

$$v(0) = v_1 + f_2$$

where v_1 is in L_2 and f_2 in $D_+(b)$. Substituting this expression for $v(0)$ into (2.36) we get

(2.37) $$v(t) = U(t)v(0) = U(t)v_1 + U(t)f_2 .$$

Since f_2 is in $D_+(b)$ we have for $t > 0$,

$$U(t)f_2 = U_0(t)f_2 = 0$$

for $|x| < b + c_{\min}t$. If v_1 were orthogonal to the nullspace of A then, by Corollary 2.1, $U(t)v_1$ would decay in the local energy norm. If not, $U(t)v_1$ and hence $U(t)v(0)$ tends locally to the projection w of v_1 on the nullspace of A. This completes the proof of Theorem 2.4.

Theorem 2.3 has the following operator theoretic interpretation: Define the operator A_+ so that its domain consists of all functions f such that

(1) f is locally in $D(A)$; that is, for every function ϕ in $C_0^\infty(R_n)$, ϕf belongs to $D(A)$ as originally defined,

(2) f is eventually outgoing, and set

$$A_+ f = A f \qquad \text{locally on} \quad D(A_+) .$$

Then for μ in the resolvent set of B^ρ, $A_+ - \mu$ is one-to-one and its range includes all functions g in H' which have compact support.

Since B^ρ has a pure point spectrum, the next result complements this assertion.

Theorem 2.5.   *The operators A_+ and B^ρ have the same point spectrum; that is, $A_+ - \mu$ has a nontrivial nullspace if and only if $B^\rho - \mu$ has a nontrivial nullspace.*

The proof of this theorem is the same as that of Theorem 4.1 in Chapter V.

One consequence of Theorem 2.2 is that B^ρ has no purely imaginary μ in its spectrum. Therefore it follows from Theorem 2.5 that no purely imaginary nonzero μ is an eigenvalue of A_+. This is a direct generalization of the *Rellich uniqueness theorem*.

Another consequence of Theorem 2.2 is that the spectrum of B^ρ is a discrete point set and by Theorem 2.5 the same thing holds for the spectrum of A_+. Further, by the spectral theory of semigroups

$$\exp (t\sigma[B^\rho]) \subset \sigma[Z^\rho(t)] ,$$

where $\sigma[K]$ denotes the spectrum of K. Therefore, if we can show that $| Z^\rho(t) |$ is less than one for some t, it follows that the spectrum of B^ρ is confined to some half-space $\text{Re } \mu < -\kappa < 0$. Likewise, if we knew that $Z^\rho(t)$ were compact for some value of t it would follow that each half-plane $\text{Re } \mu > \text{const}$ contains only a finite number of points of the spectrum of B^ρ (and hence of A_+).

Just as in the case of the acoustic equation the compactness of $Z^\rho(t)$ would follow from the *generalized Huygens' principle* provided that t is larger than the time spent by any ray in the ball of radius ρ. At present we know (Courant and Lax [1], Lax [1], Ludwig [1] that the generalized Huygens' principle is valid for the pure initial value problem. There is no doubt in the authors' minds about its validity for the mixed problem; to prove it remains an interesting technical challenge.

In Chapter V we showed that if the obstacle is star-shaped then $| Z^\rho(t) | < 1$ for t sufficiently large. The essential ingredient of that demon-

stration is a nonstandard energy identity for the wave equation due to
Morawetz and described by her in Appendix 3. She proves there that the
existence of such an identity is due to the invariance of the wave equation
under the Kelvin transformation; since general hyperbolic equations are
unlikely to remain invariant under nontrivial transformations, there is
little hope that this method can be extended.

We conclude this study of symmetric hyperbolic systems by giving a
fairly explicit description of the outgoing and incoming spectral representations for $\{U'(t)\}$, and of the scattering matrix relating the two. We start
with the free space true spectral representation

$$f \leftrightarrow \tilde{f}_0$$

which is readily obtained by a change of scale from the pre-spectral representation constructed in section one:

$$(2.38) \qquad \tilde{f}_0(\sigma, \omega) = \{(f, \phi_0(\sigma, \omega, j)) ; j = 1, \cdots, k/2\},$$

where

$$(2.39) \quad \phi_0(x; \sigma, \omega, j) = \sigma^{(n-1)/2}(2\pi\tau_j(\omega))^{-n/2} \exp\left(-i\sigma x\omega/\tau_j(\omega)\right)r_j(\omega).$$

Notice that ϕ_0 is an eigenfunction of A_0, that is, $(A_0 - i\sigma)\phi_0 = 0$. We
shall show below that the perturbed representations can be similarly
described as the inner product of f with an eigenfunction of A obtained by
adding a correction term v to ϕ_0.

In order to line up the perturbed with the unperturbed spectral representation we multiply the former by $\exp(\pm i\sigma\rho/c_{\min})$. Also, for convenience,
we assume that the nullspace of A consists only of the zero function; the
modifications necessary in the more general case are fairly trivial.

Theorem 2.6. *For real σ and ω in S_{n-1} there exist functions $v_+(x; \sigma, \omega, j)$
and $v_-(x; \sigma, \omega, j)$ which are the eventually outgoing, respectively, initially incoming solutions of*

$$(A - i\sigma)(v + \phi_0(\sigma, \omega, j)) = 0,$$

such that $v + \phi_0$ satisfies the boundary condition (2.3). We set

$$(2.40) \qquad \phi_+ = v_+ + \phi_0 \qquad and \qquad \phi_- = v_- + \phi_0$$

and define $\tilde{f}_\pm$ by

$(2.41)_+$ $\quad$ $\tilde{f}_+(\sigma, \omega) = \{(f, \phi_-(\sigma, \omega, j) ; j = 1, \cdots, k/2\}$,

$(2.41)_-$ $\quad$ $\tilde{f}_-(\sigma, \omega) = \{(f, \phi_+(\sigma, \omega, j) ; j = 1, \cdots, k/2\}$.

$\quad$ *Assertion: The correspondences*

$$f \leftrightarrow \tilde{f}_+ \quad and \quad f \leftrightarrow \tilde{f}_-$$

define outgoing and incoming spectral representations for $\{U'(t)\}$.

$\quad$ *Proof.* Let ζ be a C^∞ scalar function which is identically one near the boundary of G and vanishes outside the ball $\{|x| < \rho\}$. Set

$$w = v + \zeta\phi_0 ;$$

the equation to be satisfied by w is

$$(A - i\sigma)w = -(A - i\sigma)(1 - \zeta)\phi_0 .$$

Since A equals A_0 for $|x| > \rho$ and since ϕ_0 is annihilated by $(A_0 - i\sigma)$, it follows that the right side of this equation has its support in the ball $\{|x| < \rho\}$. The boundary condition which w has to satisfy is homogeneous. Therefore, according to Theorem 2.3 there exist two uniquely determined solutions w_+ and w_-, the first eventually outgoing and the other initially incoming.

$\quad$ The functions $\{\tau_j(\omega)\}$ and $\{r_j(\omega)\}$ can be chosen to be piecewise continuous; consequently, the functions $\phi_0(\sigma, \omega, j)$ and hence also the functions $v_\pm(\sigma, \omega, j)$ depend piecewise continuously on σ and ω in the L_2 norm over any given bounded subset.

$\quad$ The rest of the proof is exactly the same as in the proof of Theorem 5.3 of Chapter V.

$\quad$ Finally, we consider the scattering matrix for this problem. Exploiting once more the fact that the coefficients of $A - A_0$ have support in the ball $\{|x| < \rho\}$, we see that the function $v_-(x; \sigma, \omega, j)$ satisfies an equation of the form

$$(A_0 + i\sigma)v_- = g ,$$

where g is square integrable and has its support in the ball $\{|x| < \rho\}$. According to Theorem 1.6, $v_-(x; \sigma, \omega, j)$ has a free space true translation representation $k_-(s, \theta; \sigma, \omega, j)$ which for $s < -\rho/c_{\min}$ is exponential:

(2.42) $\quad\quad$ $e^{-i\sigma s}\{n_l(\theta; \sigma, \omega, j) ; l = 1, \cdots, k/2\}$.

Theorem 2.7. *The scattering matrix* $S(\sigma)$ *equals the identity plus an integral operator whose kernel K is the $k/2 \times k/2$ matrix-valued function*

$$(2.43) \qquad K_{lj}(\theta\,,\,\omega\,;\,\sigma) \;=\; (-i)^{\,(n-1)/2}(2\pi)^{1/2}\bar{n}_l(\theta\,;\,\sigma\,,\,\omega\,,\,j)$$

appearing in (2.42); *that is*

$$(2.44) \qquad
\begin{aligned}
\tilde{f}_+(\sigma,\,\omega;\,j) \;=\; & \tilde{f}_-(\sigma,\,\omega;\,j) \\
& + (-i)^{\,(n-1)/2}(2\pi)^{1/2}\sum_{l=1}^{k/2}\int_{|\theta|=1} \bar{n}_l(\theta;\,\sigma,\,\omega,\,j)\,\tilde{f}_-(\sigma,\,\theta;\,l)\;d\theta\,.
\end{aligned}$$

Proof.   The proof is essentially the same as that of Theorem 5.4, Chapter V. In order to verify (2.44) we substitute in it the expressions $(2.41)_+$ and $(2.41)_-$ for $\tilde{f}_+$ and $\tilde{f}_-$, and replace $\bar{n}_l(\,j)$ by K_{lj} ; we then obtain

$$(2.45) \qquad
\begin{aligned}
(f,\,\phi_-(\sigma,\,\omega,\,j)) \;=\; & (f,\,\phi_+(\sigma,\,\omega,\,j)) \\
& + \sum_{l=1}^{k/2}\int (f,\,\phi_+(\sigma,\,\theta,\,l)\,K_{lj}(\theta,\,\omega;\,\sigma)\;d\theta\,.
\end{aligned}$$

Interchanging the order of x and θ integration, we conclude that (2.45) holds if and only if the quantity

$$(2.46) \qquad
\begin{aligned}
\phi(x;\,\sigma,\,\omega,\,j) \;\equiv\; & \phi_-(x;\,\sigma,\,\omega,\,j) \;-\; \phi_+(x;\,\sigma,\,\omega,\,j) \\
& - \sum_{l=1}^{k/2}\int \phi_+(x;\,\sigma,\,\theta,\,l)\,\bar{K}_{lj}(\theta,\,\omega;\,\sigma)\;d\theta
\end{aligned}$$

is zero.

It therefore suffices to show that ϕ is zero. We note that each term of the sum defining ϕ is annihilated by $(A - i\sigma)$, and therefore so is ϕ. Next we show ϕ is eventually outgoing if K is defined as in (2.43). Since $i\sigma$ belongs to the resolvent set of B^ρ it then follows by Theorem 2.5 that ϕ vanishes identically.

We shall prove that ϕ is eventually outgoing by verifying that its free space true translation representer is zero for $s < -\rho/c_{\min}$. The true translation representer of ϕ is readily computed: Using the explicit formulas (1.24), (1.25), and (1.44) it is easily seen that the true translation representer $\{k_l(s,\,\theta;\,\theta',\,j);\,l = 1,\,\cdots,\,k/2\}$ of $\phi_0(x;\,\sigma,\,\theta,\,j)$ given by (2.39) is

$$(2.47) \qquad (-i)^{\,(n-1)/2}(2\pi)^{-1/2}\delta_{lj}\delta(\theta - \theta')\,e^{-i\sigma s}\,.$$

Now ϕ_+ differs from ϕ_0 by the eventually outgoing function v_+ ; consequently the value of its true translation representer on the interval $(-\infty, -\rho/c_{\min})$ is the same as that of ϕ_0. On the other hand, ϕ_- equals $v_- + \phi_0$ and hence, the value of its true translation representer on $(-\infty, -\rho/c_{\min})$ is the sum of (2.42) and (2.47). Substituting these into (2.46) we see that the value of the true translation representer of ϕ on $(-\infty, -\rho/c_{\min})$ is $e^{-i\sigma s}$ times

$$n_l(\theta' \; ; \sigma, \omega, j) - (-i)^{(n-1)/2}(2\pi)^{-1/2}\tilde{K}_{lj}(\theta', \omega ; \sigma) ,$$

which is indeed zero if K is chosen to satisfy (2.43). This completes the proof of Theorem 2.7.

Remark. The requirement that functions be eventually outgoing plays the role in these problems of a *boundary condition at infinity;* it is a version of the *Sommerfeld radiation condition.* We feel and hope that the foregoing discussion bears out our contention that of all the various formulations of the radiation condition ours is the most intuitive as well as the one which is easiest to handle technically.

PART 2

THE ACOUSTIC EQUATION WITH AN INDEFINITE ENERGY FORM AND THE SCHRÖDINGER EQUATION

3. Scattering for the Acoustic Equation with an Indefinite Energy Form

We consider the acoustic equation with a potential q which is zero outside of a compact set:

$$(3.1) \qquad u_{tt} = \Delta u - qu ,$$

over a domain G in R_3 exterior to a bounded obstacle, on the boundary of which u satisfies a condition of the form

$$(3.2) \qquad u_n + \sigma u = 0 ,$$

u_n denoting the derivative of u in the direction of the outward normal. The functions q and σ are taken to be real-valued.

Multiplying (3.1) by u_t and integrating by parts we can deduce that the 'energy':

$$(3.3) \qquad \frac{1}{2} \int_G \left[|\, u_t\,|^2 + |\, \partial_x u\,|^2 + q\,|\,u\,|^2 \right] dx + \frac{1}{2} \int_{\partial G} \sigma\,|\,u\,|^2 \, dS$$

is independent of t. In case this energy is positive definite, as would be the case when both q and σ are positive, a scattering theory can be developed along the same lines as that of Chapter V where the acoustic equation with the boundary condition $u = 0$ and without a potential was studied. If, as will happen when q and σ take on large enough negative values, the energy form (3.3) is indefinite then our former procedure is no longer applicable; nevertheless the basic ideas can be adapted so as to provide a solution to this problem.

Our main results are as follows:

I.   It is possible to define a scattering matrix $S(z)$ for the problem (3.1)–(3.2). This scattering matrix is unitary on the real axis and meromorphic in the complex plane; there may be infinitely many poles in the upper half-plane but only a finite number of poles in the lower half-plane and all of these lie on the imaginary axis.

II.   The Heisenberg scattering matrix $S^S(z)$ for the Schrödinger equation:

$$(3.4) \qquad\qquad iu_t = \Delta u - qu$$

subject to the boundary condition (3.2) is related to the acoustic scattering matrix $S(z)$ by

$$(3.5) \qquad\qquad S^S(z) = S(\sqrt{z})\ ;$$

the poles of S^S on the negative real axis which come from the poles of S in the lower half-plane are located at the energies of the bound states of the Schrödinger operator, that is, at the negative eigenvalues, if any, of $-\Delta + q$.

We start with the initial value problem for (3.1)–(3.2): Given a pair of functions $f = \{f_1, f_2\}$, to find a function $u(x, t)$ with initial values f:

$$(3.6) \qquad u(x, 0) = f_1(x) \qquad \text{and} \qquad u_t(x, 0) = f_2(x)\,,$$

which satisfies the differential equation (3.1) and the boundary condition (3.2). The differential equation (3.1) can be written symbolically in the form

$$(3.1)' \qquad\qquad u_{tt} + Lu = 0$$

where L stands for the operator

$$(3.7) \qquad\qquad L = -\Delta + q \,.$$

We prefer to think of the solution as an operator $U(t)$ relating initial data to data at time t. This suggests that we write (3.1)$'$ as a system of equations:

$$u_t = v$$

$$v_t = -Lu \,,$$

which in matrix notation is

$$(3.8) \qquad\qquad \partial_t \begin{pmatrix} u \\ v \end{pmatrix} = A \begin{pmatrix} u \\ v \end{pmatrix},$$

where

$$(3.9) \qquad\qquad A = \begin{pmatrix} 0 & I \\ -L & 0 \end{pmatrix}.$$

Denoting the initial data by f as before, we can write $\{u, v\} = U(t)f$ and express (3.8) in the form

$$(3.8)' \qquad\qquad dU(t)f/dt = AU(t)f \,.$$

To put all this on a rigorous basis it suffices to show that A can be suitably defined so that A and $-A$ satisfy the Hille–Yosida criterion for a generator of a semigroup of operators, in which case $\{U(t)\}$ is a strongly continuous group of bounded operators generated by A. We will not actually do this; we achieve the same result by showing that the space of initial data can be decomposed as a sum of subspaces P and H' each reducing A, where P is finite dimensional and where H' is such that A is skew self-adjoint on H'. Thus A generates a group of unitary operators on H' and since P is finite dimensional it generates a group of operators on P.

We introduce the customary norm $|f|_H$ on the set of initial data:

$$|f|_{H^2} = \frac{1}{2} \int_G \big[\, |\partial_x f_1|^2 + |f_2|^2 \,\big] \, dx$$

and denote by H the completion in the H-norm of smooth data with bounded support.† In addition we denote by $(\ ,\ )$ the L_2 inner product over G.

We then define A in the weak sense: Data f in H belongs to $D(A)$ if and only if there is an h in H such that

$$-(f_1, L\phi) = (h_2, \phi)$$

for all smooth functions ϕ with bounded support satisfying the boundary condition (3.2), and $f_2 = h_1$. In this case $Af = h$.

It is clear that $D(A)$ is dense in H. Using techniques from the theory of partial differential equations as in Section 2 of Chapter V, it is also not hard to prove

Lemma 3.1. *The subset*

$$F = [f \text{ in } D(A)\ ;\ |f|_H + |Af|_H \leq \text{const}]$$

is compact in the local norm:

(3.10) $$\sup_{G'} |f_1(x)| + \left\{ \int_{G'} [|\partial_x f_1|^2 + |f_2|^2]\, dx \right\}^{1/2},$$

for every bounded subdomain G' of G.

Lemma 3.2. *The operator L, weakly defined, mapping L_2 into L_2 is self-adjoint.*

The energy defined by (3.3) also plays an important role. We denote the associated bilinear form, called the *energy form*, by $(\ ,\ )_E$:

(3.11) $$(f, g)_E = \frac{1}{2} \int_G [\partial_x f_1 \overline{\partial_x g_1} + q f_1 \bar{g}_1 + f_2 \bar{g}_2]\, dx + \frac{1}{2} \int_{\partial G} \sigma f_1 \bar{g}_1\, dS.$$

For q square integrable with bounded support and σ bounded it can be shown that the energy form is a continuous Hermitian symmetric form and that A is skew symmetric relative to this form; that is, for f and g

---

† If the boundary condition (3.2) is replaced by $u = 0$, then H is defined as the completion of smooth data with bounded support satisfying the boundary condition.

in $D(A)$

$$(Af, g)_E = \frac{1}{2} \int_G [\partial_x f_2 \, \overline{\partial_x g_1} + q f_2 \bar{g}_1 - \partial_x f_1 \, \overline{\partial_x g_2} - q f_1 \bar{g}_2] \, dx$$

(3.12)
$$+ \frac{1}{2} \int_{\partial G} \sigma [f_2 \bar{g}_1 - f_1 \bar{g}_2] \, dS = -(f, Ag)_E .$$

Actually, A is almost skew self-adjoint relative to the energy form; more precisely

Lemma 3.3. *If*
$$(f, Ag)_E = (h, g)_E$$
for all g in $D(A)$, then f belongs to $D(A)$ and $Af = -h$ plus a vector in the null space of A.

Thus A is skew self-adjoint on the quotient space of H over the null space of A. Rather than work with this quotient space and have to deal with the concomitant complications (see Lax and Phillips [6]), we prefer to simplify the exposition by now assuming that zero is not an eigenvalue of A; in this case A is skew self-adjoint relative to the energy form.

One would like to say that A generates a group of unitary operators relative to the E-norm. Unfortunately, this requires that the energy form be positive definite, which, as we have mentioned earlier, need not be the case. However, the energy is positive definite on a subspace of finite codimension. To prove this we first rewrite the energy form as

(3.11)' $$(f, f)_E = (f, f)_H + Q(f),$$

where

(3.13) $$Q(f) = \frac{1}{2} \int_G q \, |f_1|^2 \, dx + \frac{1}{2} \int_{\partial G} \sigma \, |f_1|^2 \, dS .$$

The following result is not hard to prove:

Lemma 3.4. *Suppose that the potential q is square integrable with compact support and σ is bounded. Then the form $Q(f)$ is compact with respect to the form $|f|_H^2$.*

It follows from the compactness of the Q-form that

$$|Q(f)| < \tfrac{1}{2} |f|_H^2$$

for all f in H satisfying a finite number of linear conditions. Hence by (3.11)′

$$(f, f)_E \geq \tfrac{1}{2} |f|_H{}^2$$

for all such f. Now for f_1 in the domain of L restricted to L_2 both $f = \{f_1, 0\}$ and $g = \{0, f_1\}$ lie in $D(A)$; substituting these into (3.12) gives

$$(Lf_1, f_1) = \int_G \left[|\partial_x f_1|^2 + q |f_1|^2 \right] dx + \int_{\partial G} \sigma |f_1|^2 \, dS = 2(f, f)_E .$$

Consequently,

$$(Lf_1, f_1) \geq \frac{1}{2} \int_G |\partial_x f_1|^2 \, dx$$

for all f_1 in the domain of the restricted L satisfying a finite number of linear conditions. According to well-known facts of spectral theory it now follows that the restricted L *has at most a finite number of nonpositive eigenvalues*, and that (Lf_1, f_1) is positive on the orthogonal complement of the corresponding eigenfunctions.

Since we have assumed that zero is not an eigenvalue of A, it is also not an eigenvalue of L. Let $\{-\mu_j{}^2; j = 1, 2, \cdots, m\}$ denote the negative eigenvalues of L, if any, and let $\{\phi_j\}$ be the corresponding eigenfunctions:

(3.14) $$L\phi_j = -\mu_j{}^2 \phi_j .$$

It is convenient to take each μ_j to be positive.

The square of the operator A, given by (3.9), is simply

$$A^2 = \begin{pmatrix} -L & 0 \\ 0 & -L \end{pmatrix} .$$

Hence, if μ^2 is an eigenvalue of $-L$ we expect μ or $-\mu$ to be an eigenvalue of A. Indeed both are; as one may verify immediately

(3.15) $$f_j{}^+ = \{\phi_j, \mu_j\phi_j\} \quad \text{and} \quad f_j{}^- = \{-\phi_j, \mu_j\phi_j\}$$

are eigendata of A:

(3.16) $$Af_j{}^+ = \mu_j f_j{}^+ \quad \text{and} \quad Af_j{}^- = -\mu_j f_j{}^- .$$

Lemma 3.5. *The eigendata $\{f_j\}$ satisfy the following orthogonality and biorthogonality relations:*

$(3.17)_+$ $\qquad\qquad (f_j{}^+, f_k{}^+)_E = 0 \qquad$ *for all j, k*

including $j = k$;

$(3.17)_-$ $\qquad\qquad (f_j{}^-, f_k{}^-)_E = 0 \qquad$ *for all j, k*

including $j = k$;

$(3.17)_\pm$ $\qquad\qquad (f_j{}^+, f_k{}^-)_E = 0 \qquad$ *for all $j \neq k$;*

but

(3.17) $\qquad\qquad\qquad (f_j{}^+, f_j{}^-)_E \neq 0 .$

Proof. Using the skew symmetry (3.12) of A, the eigendata relations (3.16) and the fact that the numbers μ_j are real we get

$$\mu_j(f_j{}^+, f_k{}^+)_E = (Af_j{}^+, f_k{}^+)_E = -(f_j{}^+, Af_k{}^+)_E = -\mu_k(f_j{}^+, f_k{}^+)_E .$$

Since the numbers μ_j, μ_k are positive $(3.17)_+$ follows; $(3.17)_-$ can be deduced similarly. The same identities give

$$\mu_j(f_j{}^+, f_k{}^-)_E = \mu_k(f_j{}^+, f_k{}^-)_E ,$$

for which $(3.17)_\pm$ follows if $\mu_j \neq \mu_k$.

The verification of (3.17) makes use of the definition (3.15) of $f_j{}^\pm$, the form (3.11) of the energy inner product, the expression (3.12) and the eigenfunction equation (3.14):

$$(f_j{}^+, f_j{}^-)_E = \tfrac{1}{2}[\mu_j{}^2(\phi_j, \phi_j) - (\phi_j, L\phi_j)] = \mu_j{}^2(\phi_j, \phi_j) ,$$

which is indeed different from zero.

Finally, suppose that an eigenvalue $-\mu^2$ of L has multiplicity greater than one. Then there are a corresponding number of eigendata f^+ and f^- of A with eigenvalues μ and $-\mu$, respectively. It follows from (3.17) that we may set up biorthogonal sets in the null-spaces of $A - \mu$ and $A + \mu$ so that the relation $(3.17)_\pm$ is satisfied even when $\mu_j = \mu_k$.

Let P denote the subspace spanned by the eigenfunctions $\{f_j{}^+, f_j{}^-\}$.

Corollary 3.1. *The energy form is nondegenerate on P; that is, only the zero element in P is E-orthogonal to all of P.*

Next let H' denote the set of all data in H which are E-orthogonal to P.

Corollary 3.2. *Every element f of H has a unique E-projection into H'; that is, every f has a unique decomposition of the form*

$$f = g + p$$

where g lies in H' and p lies in P.

Corollary 3.3. *A maps P into P and H' into H'.*

Proof. It is obvious that P is an invariant subspace of A and since A is skew symmetric so is the E-orthogonal complement H'.

Lemma 3.6. *The energy form is positive definite on H' and defines an equivalent metric on H'; that is, there exist positive constants c and C such that*

$$(3.18) \qquad c(f,f)_E \le |f|_{H^2} \le C(f,f)_E$$

for all f in H'.

Proof. We first prove that the energy form is positive on H'. It is clear from the definition of the eigendata given in (3.15) that if f belongs to H' then so does $h = \{f_1, 0\}$ and further it is clear from the definition of the energy form that

$$(h, h)_E \le (f, f)_E .$$

Consequently, the positivity of the E-form on H' will be established if we can show that $(h, h)_E \ge 0$. To this end let

$$h_k = \{u_k, 0\}$$

be a sequence of smooth data with bounded support satisfying the boundary condition (3.2) and converging to h in the H-norm. An integration by parts gives

$$(h_k, f_j^+)_E = \tfrac{1}{2}(u_k, L\phi_j) = -\frac{\mu_j^2}{2}(u_k, \phi_j)$$

and since h_k tends to h and h is by assumption orthogonal to f_j^+, we see that

$$(3.19) \qquad \lim_{k \to \infty} (u_k, \phi_j) = 0 .$$

Let u_k' be the L_2-orthogonal projection of u_k into the L_2-orthogonal complement of the negative eigenspace of L. Since this eigenspace is finite dimensional u_k' belongs to the domain of L restricted to L_2 and further u_k' belongs to the nonnegative eigenspace of L. Hence, for $h_k' = \{u_k', 0\}$

$$(h_k', h_k')_E = (Lu_k', u_k') \geq 0 .$$

On the other hand because of (3.19), $\{h_k'\}$ converges with $\{h_k\}$ to h in the H-norm and therefore

$$(h, h)_E = \lim (h_k', h_k')_E = \lim (Lu_k', u_k') \geq 0 .$$

Next we establish the relation (3.18). The left inequality in (3.18) follows from the continuity of the E-form in the H-norm. If the right inequality were not true then, because of the positivity of the E-form on H', there would exist a sequence $\{f_n\}$ in H' for which

(3.20) $$\lim (f_n, f_n)_E = 0 \qquad \text{and} \qquad |f_n|_H = 1 ;$$

we may as well suppose that this sequence converges weakly in H' to data f_0. Since the Q-form is compact by Lemma 3.4 we can assert that

$$Q(f_0) = \lim Q(f_n) .$$

Now according to (3.11)$'$

$$(f_n, f_n)_E = |f_n|_{H}^2 + Q(f_n)$$

and since

$$|f_0|_H \leq \lim \inf |f_n|_H$$

we see from (3.20) that

$$Q(f_0) = -1 \qquad \text{and} \qquad (f_0, f_0)_E \leq 0 .$$

Hence f_0 is a nonzero element and since the E-form is positive on H' we conclude that $(f_0, f_0)_E = 0$ and therefore by the Schwarz inequality that $(f_0, g)_E = 0$ for all g in H'. On the other hand f_0 is by construction orthogonal to P so that $(f_0, g)_E = 0$ for all g in H. In particular for all g in $D(A)$ we have $(f_0, Ag)_E = 0$ and it follows from Lemma 3.3 that f_0 belongs to $D(A)$ and $Af_0 = 0$, contrary to our assumption that zero is not an eigenvalue for A. This completes the proof of Lemma 3.6.

We can now discuss the solution to the initial value problem posed by (3.1) and (3.2).

Lemma 3.7. *The operator A generates a strongly continuous group of operators $\{U(t)\}$ on H. These operators leave H' invariant and restricted to*

H' they form a group of unitary operators in the energy norm. The following inequality holds:

(3.21) $E(U(T)f, R) \le E(f, R + T)$

where

$$E(f, R) = \int_{G(R)} \left[|\, \partial_x f_1 \,|^2 + q\,|\, f_1 \,|^2 + |\, f_2 \,|^2 \right] dx + \int_{\partial G} \sigma\,|\, f_1 \,|^2\, dS$$

and $G(R)$ denotes the set of points common to G and the ball $\{|\, x \,| < R\}$.

Proof. The action of $U(t)$ on P is determined by (3.16):

$$U(t)\left[\sum a_j f_j^+ + \sum b_j f_j^-\right] = \sum a_j \exp(\mu_j t)\, f_j^+ + \sum b_j \exp(-\mu_j t)\, f_j^- ;$$

whereas $\{U(t)\}$ acts on H' like a group of unitary operators in the energy norm and, because of the equivalence of metrics, like a bounded group of operators on H' in the H-norm. Since P is finite dimensional the projections of H on P and H' are bounded in the H-norm and therefore the effect of $\{U(t)\}$ on H itself is that of a strongly continuous group of bounded operators with exponential growth in $|\, t \,|$. The inequality (3.21) can be established by the same argument used in the proof of Theorem 1.3 of Chapter V.

In scattering theory one always works in the complement of the eigen-space and we therefore restrict the discussion to the subspace H' where $\{U(t)\}$ forms a one-parameter group of operators, unitary with respect to the energy form. Unless otherwise stated we will from now on use only the E-norm on H'. We now have to define incoming and outgoing subspaces for this group of operators.

If H' were equal to H we would define the incoming and outgoing sub-spaces as before: A solution u of (3.1)–(3.2) is called incoming [outgoing] if it vanishes in the cone $\{|\, x \,| < \rho - t;\, t < 0\}$ $[\{|\, x \,| < \rho + t;\, t > 0\}$, where ρ is chosen so that both the obstacle and the support of the potential q are contained in the ball $\{|\, x \,| < \rho\}$. Then D_- and D_+ consist of data for which the corresponding solution of (3.1)–(3.2) is incoming and outgoing, respectively.

In general H' will be a proper subspace of H and in this case D_- and D_+ defined as above do not lie in H'; to overcome this difficulty we have to find subspaces in H' which play the role of D_- and D_+ . Two possibilities immediately suggest themselves and both turn out to be essential to our discussion.

Definition 3.1. D_-' and D_+' are the E-projections of D_- and D_+, respectively, into H'.

Definition 3.2. D_-'' and D_+'' are the intersections of D_- and D_+, respectively, with H'.

Theorem 3.1. D_-' and D_-'' are incoming subspaces and D_+' and D_+'' are outgoing subspaces of H' for the group $\{U(t)\}$.

Proof. We recall (see Chapter II) that a closed subspace D of H' is outgoing if it has the following three properties:

(i) $U(t)D \subset D$ for $t > 0$,
(ii) $\cap \, U(t)D = \{0\}$,
(iii) $\cup \, U(t)D$ is dense in H'.

To show that D_+' has property (i) we take g in D_+'. By Corollary 3.2 and the definition of D_+' such a g has a unique† decomposition of the form

$$(3.22) \qquad g = f + p, \qquad f \text{ in } D_+ \text{ and } p \text{ in } P.$$

Conversely, every g in H' of the form (3.22) lies in D_+'. Applying $U(t)$ to (3.22) we get

$$(3.22)_t \qquad\qquad U(t)g = U(t)f + U(t)p \equiv f_t + p_t.$$

Since H' and P are invariant under $U(t)$ as is D_+ for $t > 0$, the decomposition $(3.22)_t$ of $U(t)g$ is of the type (3.22) and hence $U(t)g$ belongs to D_+' for $t > 0$.

That D_+'' has property (i) is trivial since D_+'' is the intersection of two sets, D_+ and H', both of which have property (i).

Next we show that D_+' has property (ii); suppose in fact that g belongs to $U(t)D_+'$ for all t. Such a g can be written as

$$g = U(t)h_t,$$

where h_t belongs to D_+' and hence has the decomposition

$$h_t = f_t + p_t, \qquad f_t \text{ in } D_+ \text{ and } p_t \text{ in } P.$$

† Clearly $D_+ \cap P = \{0\}$.

Combining these relations we get

$$(3.23)_t \qquad\qquad g = U(t)f_t + U(t)p_t .$$

Now $U(t)$ maps P into itself and hence the second term in $(3.23)_t$ belongs to P. On the other hand f_t belongs to D_+ and therefore the first term vanishes in the ball $\{|x| < \rho + t\}$. Since t is arbitrary $(3.23)_t$ show that in each ball g is equal to some element of P and since P is finite dimensional it follows from this that g itself belongs to P. However, g also belongs to H' and is therefore E-orthogonal to P; according to Corollary 3.1 the only element g of P which is E-orthogonal to P is the zero element, which is what we wanted to prove.

Since D_+'' is a subspace of D_+' it has property (ii) *a fortiori* if D_+' does.

The proof of property (iii) for D_+' is the same as in the case without potential and the simpler boundary condition presented in Chapter V. The essential ingredients are all here: (a) We have assumed that zero is not an eigenvalue of A; (b) The Rellich theorem combined with the unique continuation property of the operator (which we assume) shows that A has no point spectrum other than zero as in the proof of Theorem 2.2 of Chapter V; and (c) The set of data for which $|f|_H + |Af|_H$ is bounded are precompact in the local norm (3.10).

Therefore our previous proof of property (iii) goes through for D_+'. We will indicate later how to prove property (iii) for D_+'' from the D_+'-translation representation. The incoming case can be handled similarly; this completes the proof of Theorem 3.1.

Having proved the incoming and outgoing character of D_-', D_-'', D_+', D_+'' we can, according to the translation representation theorem, associate a unitary translation representation with each of these spaces. We shall denote these representers as k_-', k_-'', k_+', k_+'', respectively.

An operator relating any two representers of the same f is a kind of scattering operator. However the one which arises in the customary fashion from the wave operators relates k_-' and k_+':

$$(3.24) \qquad\qquad S: \; k_-' \rightarrow k_+' .$$

To see this we take as the wave operators

$$(3.25) \qquad\qquad W_\pm f = \lim_{t\to\pm\infty} U(-t)P'P_0 U_0(t)f ;$$

P' and P_0 are defined on page 237. For any f in D_-, $g = P'P_0 f$ belongs† to D_-' and can be expressed as in (3.22):

$$(3.22)_- \qquad\qquad g = f + p, \qquad p \quad \text{in} \quad P.$$

For $t < 0$ it is clear that $U_0(t)f = U(t)f$ and hence that

$$U(t)g = U_0(t)f + U(t)p.$$

This is again a decomposition of the type (3.22); therefore, $P'P_0U_0(t)f = U(t)g$ and

$$(3.26)_- \qquad\qquad W_-f = g.$$

Thus $W_-D_- = D_-'$ and this together with the fact that D_- and D_-' are incoming subspaces implies that W_- exists, is unitary‡ on H_0 to H' (in the energy norm) and can be represented as taking $W_-^{-1}h$ with free space translation representer k_0 into h with D_-'-translation representer $k_-' = k_0$ (see Section 2 of Chapter I). A similar assertion holds for W_+ and it follows that $S = W_+^{-1}W_-$ is represented by (3.24).

To study the properties of S we factor the mapping $k_-' \to k_+'$ as follows:

$$k_-' \to k_-'' \to k_+'' \to k_+'.$$

We introduce the following notation for the scattering operators which occur in the above decomposition:

$$
\begin{aligned}
S_- &: \quad k_-'' \to k_-' \\
(3.27) \qquad\qquad S_+ &: \quad k_+' \to k_+'' \\
S'' &: \quad k_-'' \to k_+''.
\end{aligned}
$$

In terms of these we can write

$$(3.28) \qquad\qquad S = S_+^{-1}S''S_-^{-1}.$$

Lemma 3.8. *The three operators S_-, S_+ and S'' are causal; that is, they map functions k whose support lies in $(-\infty, 0)$ into functions of the same kind.*

---

† P' and the E-projection coincide on functions which vanish in $\{|x| < \rho\}$.

‡ To prove that W_- is an isometry it is enough to show in the decomposition $(3.22)_-$ that $|g|_E = |f|_{H_0}$; that $|g|_E = |f|_E$ follows from (3.34) and $(3.17)_-$ and since f has its support outside the ball $\{|x| < \rho\}$ we also have $|f|_E = |f|_{H_0}$.

Proof. We shall base the proof of all three statements on Lemma 4.1 of Chapter II according to which the scattering operator associated with an *orthogonal* pair of incoming and outgoing subspaces is causal. Our proof also depends on the easily verified fact that the orthogonal complement of an incoming [outgoing] subspace is an outgoing [incoming] subspace.

Now S_- is associated with the incoming subspace D_-'' and the outgoing subspace $D_-'^\perp$, where the symbol $M^\perp$ means the E-orthogonal complement of M in H'. Since D_-' contains D_-'' it follows that $D_-'^\perp$ is orthogonal to D_-'' and by the lemma quoted above S_- is causal.

In the same way S_+ is associated with the incoming and outgoing subspaces $D_+'^\perp$ and D_+'', respectively. Again since D_+' contains D_+'' it follows that $D_+'^\perp$ is orthogonal to D_+''; this proves the causality of S_+.

Finally, S'' is causal since D_-'' and D_+'', being subspaces of the orthogonal subspaces D_- and D_+ (see Corollary 2.2 of Chapter IV), are themselves H-orthogonal and hence, since their support is outside the ball $\{|x| < \rho\}$, they are also E-orthogonal.

Remark. As we shall show below in general D_-' and D_+' are *not* orthogonal and therefore the scattering operator S is not causal.

We recall that in the spectral representation, which is the Fourier transform of the translation representation, the scattering operator goes over into multiplication by the scattering matrix. According to Theorem 4.1 of Chapter II the scattering matrix-valued function associated with causal operators is the boundary value of a bounded analytic function in the lower half-plane. Corresponding to the factorization (3.28) of the scattering operator we have the following factorization of the scattering matrix:

$$(3.28)' \qquad S(\sigma) = S_+^{-1}(\sigma)S''(\sigma)S_-^{-1}(\sigma).$$

To locate the domain of analyticity of $S(z)$ we shall determine the domains of analyticity of the factors and the location of their zeros.

Lemma 3.9. S_- *and* S_+ *are each meromorphic in the whole plane with zeros at* $\{-i\mu_j ; j = 1, \cdots, m\}$ *and at no other points, and poles at* $\{i\mu_j ; j = 1, \cdots, m\}$.

Proof. We shall prove the lemma for S_-. According to Theorem 3.2 of Chapter III, $-i\mu_j$ is a zero of S_- if and only if $\exp(-u_j t)$ is an eigenvalue of the associated semigroup operator $Z(t)$:

$$Z(t) = P_+U(t)P_- \qquad (t \geq 0) ,$$

where P_- and P_+ are orthogonal projections onto the orthogonal complement of the incoming and outgoing subspaces which in this case are D_-'' and $D_-'^\perp$, respectively. The domain K of $\{Z(t)\}$ is the orthogonal complement of the incoming and outgoing subspaces; in this case

$$(3.29) \qquad\qquad K = D_-' \ominus D_-'' .$$

The above version of Theorem 3.2 of Chapter III holds when K is finite dimensional.

As in formula (3.22) every g in D_-' can be represented uniquely in the form

$$(3.22)_- \qquad\qquad g = f + p ,$$

f in D_- and p in P. Furthermore, g belongs to D_-'' if and only if $p = 0$. Thus, $(3.22)_-$ defines a linear map of D_-' into P whose kernel is D_-''; the dimension of K is therefore the same as the dimension of the range of this mapping. Since P, being spanned by the $2m$ eigendata $\{f_j^-, f_j^+\}$, is of dimension $2m$, it follows that K can be at most of dimension $2m$.

Proposition 3.1. K *is* m-*dimensional.*

Proof. We shall use the unperturbed translation representation $\mathfrak{R}$ which has been discussed in detail in Chapter IV. In particular one of the basic properties of this representation is that for every f in D_-

$$(3.30) \qquad\qquad [\mathfrak{R}f](s) = 0 \qquad \text{for} \quad s > -\rho .$$

Outside the ball of radius ρ the data f_j^+, f_j^- satisfy the equations:

$$(3.31)_+ \qquad\qquad (A_0 - \mu_j)f_j^+ = 0 \qquad (|x| > \rho) ,$$

$$(3.31)_- \qquad\qquad (A_0 + \mu_j)f_j^- = 0 \qquad (|x| > \rho) .$$

Define f_j^+ and f_j^- arbitrarily inside the obstacle so that they become free space data and hence have free space translation representers. It then follows from (3.31) by Remark 4.2 of Chapter IV that $\mathfrak{R}f_j^\pm$ is of the form

$$[\mathfrak{R}f_j^+](s) = \exp{(-\mu_j s)}n_j^+ \qquad (|s| > \rho) ,$$

and

$$[\mathfrak{R}f_j^-](s) = \exp{(-\mu_j s)}n_j^- \qquad (|s| > \rho) .$$

Since the $\Re f_j^{\pm}$ are square integrable they can contain no increasing exponential parts; consequently

$$(3.32)_+ \qquad [\Re f_j^+](s) = \begin{cases} 0 & \text{for } s < -\rho \\ \\ \exp(-\mu_j s)\, n_j^+ & \text{for } s > \rho, \end{cases}$$

$$(3.32)_- \qquad [\Re f_j^-](s) = \begin{cases} \exp(\mu_j s)\, n_j^- & \text{for } s < -\rho \\ \\ 0 & \text{for } s > \rho. \end{cases}$$

Comparing (3.30) and $(3.32)_+$ we see immediately that for f in D_- the supports of $\Re f$ and of $\Re f_j^+$ are disjoint and therefore

$$[\Re f, \Re f_j^+] = 0 .$$

Since $\Re$ is a unitary representation with respect to the unperturbed energy norm in H_0, we deduce that for f in D_-

$$(3.33) \qquad (f, f_j^+)_{H_0} = [\Re f, \Re f_j^+] = 0 .$$

Now data f in D_- vanishes inside the ball $\{|x| < \rho\}$ and therefore we may in (3.33) replace the H_0-inner product by the E-inner product and conclude that the f_j^+ are orthogonal† to D_-:

$$(3.34) \qquad (f, f_j^+)_E = 0 \qquad \text{for } f \text{ in } D_- .$$

It follows from (3.34) and Definition 3.2 that an f in D_- belongs to D_-'' if and only if

$$(3.35) \qquad (f, f_j^-)_E = 0 \qquad \text{for } j = 1, \cdots, m .$$

Since f belongs to D_- we may, as pointed out earlier, write the E-inner products as H_0-inner products; these in turn are equivalent to the inner products of the corresponding unperturbed translation representers. Therefore we can write (3.35) as

$$(3.36) \qquad [\Re f, \Re f_j^-] = 0 \qquad \text{for } j = 1, \cdots, m .$$

We denote $\Re f$ by k; then $k(s) = 0$ for $s > -\rho$ by (3.30). On the other hand according to $(3.32)_-$, $\Re f_j^-$ is exponential for $s < -\rho$. Hence we can

---

† It is not hard to give a direct proof of (3.34) without using the free space translation representation but rather the behavior of $(U(t)f, f_j^+)_E$ as t tends to $-\infty$.

rewrite condition (3.36) as

$$(3.37) \qquad \int_{-\infty}^{-\rho} k(s) \cdot n_j^- \exp(\mu_j s) \, ds = 0 \qquad \text{for} \quad j = 1, \cdots, m.$$

Denote by g the projection of f in D_- on H':

$$(3.22)_- \qquad\qquad g = f + p, \qquad p \quad \text{in} \quad P.$$

As shown in (3.26), $W_- f = g$ where W_- is the wave operator defined in (3.25). As explained in Section 2 of Chapter I, the D_-'-translation representer k_-' of g is the same (except for a shift to the right by ρ) as the unperturbed translation representer k of f. It therefore follows from (3.37) that k_-' with support on $(-\infty, 0)$ is the representer of an element in D_-'' if and only if

$$(3.38) \qquad \int_{-\infty}^{0} k_-'(s) \cdot n_j^- \exp(\mu_j s) \, ds = 0 \qquad \text{for} \quad j = 1, \cdots, m.$$

Since D_-' itself maps onto $L_2(-\infty, 0; N)$ in the D_-'-translation representation, we conclude from the above that the dimension of K is just the dimension of the set

$$(3.39) \quad k_j(s) = \begin{cases} n_j^- \exp(\mu_j s) & \text{for} \quad s < 0, \\ \\ 0 & \text{for} \quad s > 0; \end{cases} \qquad (j = 1, \cdots, m).$$

Now a set of k_j's corresponding to different eigenvalues are certainly linearly independent if none of the n_j^-'s are zero, and it is also clear that the whole set will be linearly independent if it can be shown that any f^-, for which $(A + \mu)f^- = 0$ and for which the corresponding $n^- = 0$, is itself zero. By $(3.32)_-$ $\Re f_-$ will then be zero for $|s| > \rho$. Moreover, according to Corollary 4.3 of Chapter IV, data f for which $(A_0 + \mu)f$ vanishes for $|x| > \rho$, $\mu \neq 0$, and for which $\Re f$ vanishes for $|s| > \rho$ must itself vanish for $|x| > \rho$. But then by the assumed unique continuation property of the operator A it follows that such an f is zero everywhere. This completes the proof of the assertion that K is m-dimensional.

Let f_j denote the element in H' with D_-'-translation representer k_j given in (3.39). We now complete the proof of Lemma 3.9 by showing that each f_j is an eigenelement of the semigroup operator $Z(t)$ with eigenvalue $\exp(-\mu_j t)$. It follows from (3.39) that f_j belongs to D_-' and from (3.38) that f_j is orthogonal to D_-''. Consequently, f_j lies in K and $P_- f_j = f_j$. Further, $U(t)f_j$ has as its D_-'-translation representer $k_j(s - t)$ and the action of P_+ on this is to restrict its support to $(-\infty, 0)$; the result is simply $\exp(-\mu_j t)k_j$. Hence

$$Z(t)f_j = \exp(-\mu_j t)f_j$$

as desired. Since K is m-dimensional and we have constructed m linearly independent eigenelements this completely determines the spectrum of $Z(t)$. Hence, according to Theorem 3.2 of Chapter III S_- has zeros at $\{-i\mu_j ; j = 1, \cdots, m\}$ and nowhere else. That S_- has poles at the points $\{i\mu_j\}$ follows from Theorem 5.1 of Chapter III.

The analogous result for S_+ is proved in a similar fashion from the adjoint of the semigroup associated with the subspaces D_+' and D_+''.

Corollary 3.4. $S_-^{-1}(z)$ and $S_+^{-1}(z)$ are meromorphic in the complex plane having poles at the points $\{-i\mu_j ; j = 1, \cdots, m\}$ and only these points.

Proof. This follows from the relation

$$S^{-1}(z) = S^*(\bar{z})$$

and Lemma 3.9.

We now complete the proof of Theorem 3.1.

Proposition 3.2. D_-'' and D_+'' satisfy property (iii).

Proof. It follows from (3.38) that k_-' is the D_-'-translation representer of an element in $\cup\, U(t)D_-''$ if and only if $k_-'(s)$ vanishes for sufficiently positive s and

$$\int_{-\infty}^{\infty} k_-'(s) \cdot n_j^- \exp(\mu_j s)\, ds = 0 \qquad \text{for} \quad j = 1, \cdots, m.$$

Since the k_j are linearly independent, no nontrivial linear combination of the functions defined as

(3.40) $n_j^- \exp(\mu_j s)$ on $(-\infty, \infty)$ $(j = 1, \cdots, m)$,

is square integrable. By a well-known theorem the set of k_-', which are of bounded support and orthogonal to the functions (3.40), are dense in $L_2(-\infty, \infty; N)$; this proves the D_-'' part of Proposition 3.2 and D_+'' can be treated similarly.

In order to determine the analytic behavior of S we still have to investigate the third factor S''.

Lemma 3.10. *S'' is meromorphic in the complex plane with poles only in the upper half-plane* Im $z > 0$.

Proof. Again we make use of the associated semigroup of operators, in this case

$$Z''(t) = P_+''U(t)P_-'',$$

where P_-'' and P_+'' are projections of H' onto the orthogonal complements of D_-'' and D_+'', respectively. These operators form a semigroup on the subspace

$$K'' = H' \ominus (D_-'' \oplus D_+'')$$

with infinitesimal generator B''. We shall prove that $Z''(2\rho)(\kappa I - B'')^{-1}$ is compact and the assertion of the lemma will then follow from Theorem 5.1 of Chapter III. The proof of this fact is similar to that of Theorem 3.1 of Chapter V. However, because we work in the subspace H' rather than H certain modifications in the argument are required; these are the subject of the next two propositions.

We begin by introducing the following operators:

E A linear bounded map of H into H_0 which extends data into the complement of G;

P_0 Projects H_0 onto H by simply restricting the support of data in H_0 to G;

P' The E-orthogonal projection of H onto H';

$M = P'[U(2\rho) - P_0U_0(2\rho)E]$;

P_-' and P_+' The projections of H' onto the orthogonal complements of D_-' and D_+', respectively.

Proposition 3.3. *M is a bounded linear operator and*

$$J' = P_+''P'P_0U_0(2\rho)EP_-''$$

is a compact linear operator on H'.

Proof. It is clear from their definitions that M and J' are bounded linear operators on H'. In order to prove that J' is in addition compact it suffices to show that its range is finite dimensional. Given h in H' we write

$$P_-''h = P_-'h + (P_-'' - P_-')h ,$$

where $P_-'h$ is orthogonal to D_-' by the definition of P_-'. Any f in D_- can be decomposed as in $(3.22)_-$:

$$g = f + p \qquad g \text{ in } D_-' \text{ and } p \text{ in } P .$$

Now $P_-'h$ is orthogonal to both P and D_-' so that $(P_-'h, f)_E = 0$; thus $P_-'h$ is E-orthogonal to D_-. Since the elements of D_- have their support outside the ball $\{|x| < \rho\}$ and since $P_-'h$ is equal to $EP_-'h$ on this set, we see that $EP_-'h$ will be H_0-orthogonal to D_-. According to the unperturbed translation representation theory presented in Section 2 of Chapter IV, $U_0(2\rho)EP_-'h$ belongs to D_+ and hence

$$h' \equiv P'P_0U_0(2\rho)EP_-'h = P'U_0(2\rho)EP_-'h$$

lies in D_+'. Thus $P_+'h' = 0$ and

$$P_+''h' = (P_+'' - P_+')h'$$

so that we finally have

$$J'h = P_+''P'P_0U_0(2\rho)E(P_-'' - P_-')h + (P_+'' - P_+')P'P_0U_0(2\rho)EP_-'h .$$

The range of $P_\pm'' - P_\pm'$ is finite dimensional by Proposition 3.1 and it follows from this that the range of J' is also finite dimensional.

Proposition 3.4.   *For κ sufficiently large the operator $Z''(2\rho)(\kappa I - B'')^{-1}$ is compact.*

Proof.   Choose f in K''. As in the proof of Theorem 3.1 of Chapter V the Laplace representation for $(\kappa I - B'')^{-1}$ gives

$$(3.41) \quad g = Z''(2\rho)(\kappa I - B'')^{-1}f = P_+''U(2\rho)P_-'' \int_0^\infty e^{-\kappa t}U(t)P_-''f\,dt ,$$

where the P_-'' before the integral has no effect on the integral. Therefore for κ sufficiently large (3.41) can be rewritten in terms of M, J' and the resolvent of A as

$$g = (P_+''M + J')(\kappa I - A)^{-1}P_-''f .$$

The operator J' is known to be compact by Proposition 3.3 and therefore the operator on the right in this expression will be compact if

$$J'' = P_+''M(\kappa I - A)^{-1}P_-''$$

is compact. Now $(\kappa I - A)^{-1}$ takes any bounded set into a set F such that F and AF are both bounded and hence by Lemma 3.1 the set F is compact in the local norm (3.10). On the other hand a domain of dependence argument shows that Mh depends only on the behavior of h in the ball $\{|x| < 5\rho\}$. It follows that $M(\kappa I - A)^{-1}P_-''$ is compact and so is J''. This completes the proof of Proposition 3.4 and Lemma 3.10 as well.

The culminating result of this section is

Theorem 3.2. *The scattering matrix* $\mathcal{S}$ *is meromorphic in the complex plane, holomorphic on the real axis and the only poles in the lower half-plane occur at the points* $\{-i\mu_j ; j = 1, \cdots, m\}$.

Proof. According to Lemma 3.8, $\mathcal{S}''$ is holomorphic in the lower half-plane and $\mathcal{S}_\pm^{-1}$ are holomorphic in the upper half-plane; according to Corollary 3.4 and Lemma 3.10 all of these matrix-valued functions are holomorphic on the real axis. Hence the singularities of

$$(3.28)' \qquad\qquad \mathcal{S} = \mathcal{S}_+^{-1}\mathcal{S}''\mathcal{S}_-^{-1}$$

are determined by the poles of $\mathcal{S}_\pm^{-1}$ in the lower half-plane and by those of $\mathcal{S}''$ in the upper half-plane. The assertion of the theorem therefore follows from Corollary 3.4 and Lemma 3.10 if it can be shown that none of the poles in the lower half-plane of $\mathcal{S}_+^{-1}$ and $\mathcal{S}_-^{-1}$ are cancelled out by zeros of $\mathcal{S}''$.

This latter fact can be established by a decomposition of S as the sum of two operators each of which commute with translations:

$$S = S_c + S_n,$$

where S_c is causal and S_n, the noncausal part, takes $L_2(-\infty, 0; N)$ into $L_2(0, \infty; N)$. In the corresponding spectral representation of S, namely,

$$\mathcal{S} = \mathcal{S}_c + \mathcal{S}_n,$$

$\mathcal{S}_c$ is holomorphic in the lower half-plane and $\mathcal{S}_n$ can be shown to be equal to

$$[\mathcal{S}_n \tilde{f}](\sigma) = \sum_{j=1}^m \frac{c_j}{\sigma + i\mu_j} (\tilde{f}(\sigma), n_j^-)_N n_j^+$$

where $\{n_j{}^\pm\}$ are determined by $(3.32)_\pm$; they have been shown in the proof of Proposition 3.1 to be nonzero. Since this is a rather fine point we do not give the details of the argument, but refer the reader to the original paper (see Lax and Phillips [6]).

Remark. For the pure initial-value problem (that is when $G = R_3$), if q has compact support and belongs to $L_{3+\delta}$, $\delta > 0$, then it can be shown (see Lax and Phillips [6]) that $\mathcal{S}$ has only a finite number of poles in any half-plane Im $z < c$.

4. The Schrödinger Scattering Matrix

We now show that the scattering matrix $\mathcal{S}^S$ for the Schrödinger wave equation

$$(4.1) \qquad\qquad u_t = iLu$$

is related to the acoustic scattering matrix $\mathcal{S}$ obtained in the previous section by

$$(4.2) \qquad\qquad \mathcal{S}^S(z) = \mathcal{S}(\sqrt{z}) \ .$$

Our proof of this fact is based on the following observation:

$$(4.3) \qquad\qquad A^2 = \begin{pmatrix} -L & 0 \\ 0 & -L \end{pmatrix},$$

where A defined by (3.9) is the generator for the acoustic problem. We see that A^2 acts like the Hamiltonian $-L$ on each component of the acoustic data separately; in particular it acts like $-L$ restricted to L_2 on the second component. Taking A^2 as our perturbed operator and

$$(4.4) \qquad\qquad A_0{}^2 = \begin{pmatrix} \Delta & 0 \\ 0 & \Delta \end{pmatrix}, \qquad G_0 = R_3 \ ,$$

as our unperturbed operator, the corresponding wave operators are defined as:

$$(4.5) \qquad W_\pm{}^S f = \lim_{t\to\pm\infty} \exp\,(-iA^2t)P'P_0 \exp\,(iA_0{}^2t)f \ ;$$

here P_0 projects H_0 onto H and P' projects H onto H'. Again for simplicity we shall suppose that zero is not an eigenvalue of A. Since the two components of the data are completely uncoupled by both A_0^2 and A^2, we obtain the usual Schrödinger wave operators by projecting down to the second component.

Lemma 4.1.   *The wave operators $W_\pm^S$ exist and are unitary from H_0 to H'. Moreover, if f_+ is an element of H_0 whose spectral representer with respect to A_0 has its support on the positive real axis, then*

$(4.6)_+$ $$W_\pm^S f_+ = W_\pm f_+ ;$$

similarly

$(4.6)_-$ $$W_\pm^S f_- = W_\mp f_-$$

for all f_- whose spectral representers have their support on the negative real axis.

Proof.   This result is essentially a special case of Theorem 4.2 of Chapter II with $\phi(t) = t^2$; the difference arises from the fact that we have to project down into H' in the range space. In order to prove $(4.6)_+$ it suffices to consider data f_+ whose spectral representer is smooth with compact support in $(0, \infty)$. We set

(4.7) $$\exp(iA_0^2 t)f_+ = g_1(t) + g_2(t) ,$$

with $g_1(t)$ in D_+ and $g_2(t)$ orthogonal to D_+. In Lemma 4.2 of Chapter II $\exp(iA_0^2 t)f_+$ was split into components $g_1(t)$ in D_+^0 and $g_2(t)$ orthogonal to D_+^0 where D_+^0 is the free space outgoing subspace defined for $\rho = 0$, and it was proved there that

(4.8) $$\lim_{t \to \infty} g_2(t) = 0 .$$

Actually, the same argument shows that for any decomposition of this type, $g_1(t)$ in D_+^c and $g_2(t)$ orthogonal to D_+^c, the relation (4.8) continues to hold.

Next we decompose $g_1(t)$ which lies in $D_+ = D_+^\rho$ into its projection $g_1'(t) = P'g_1(t)$ and the remainder $g_1''(t) = (I - P')g_1(t)$ in P. According to the D_+ analogue of (3.34)

$$(g_1''(t), f_j^-)_E = (g_1(t), f_j^-)_E = 0 \qquad \text{for} \quad j = 1, \cdots, m .$$

On the other hand since $g_1(t)$ belongs to D_+ the support of $g_1(t)$ lies outside the ball $\{|x| < \rho\}$ and therefore

$$(g_1''(t), f_j^+)_E = (g_1(t), f_j^+)_E = (g_1(t), f_j^+)_{H_0}.$$

Let $k(s, t)$ be the free space translation representer of $g_1(t)$:

$$k(s, t) = \Re g_1(t) = \begin{cases} \int_0^\infty \exp(-i\sigma s + i\sigma^2 t)\tilde{f}_+(\sigma)\, d\sigma & \text{for } s > \rho, \\ \\ 0 & \text{for } s < \rho; \end{cases}$$

here $\tilde{f}_+$ denotes the free space spectral representer of f_+. By $(3.32)_+$

$$[\Re f_j^+](s) = n_j^+ \exp(-\mu_j s) \qquad \text{for } s > \rho.$$

Combining these various statements we get

$$(g_1''(t), f_j^+)_E = [k(t), \Re f_j^+] = \int_\rho^\infty (k(s, t), n_j^+)_N \exp(-\mu_j s)\, ds.$$

We split the range of integration into two parts: one over (ρ, c) and the other over (c, ∞). As mentioned above, the argument used in the proof of Lemma 4.2 of Chapter II shows that

$$\lim_{t \to \infty} \int_\rho^c |k(s, t)|^2\, ds = 0,$$

and consequently the integral over (ρ, c) tends to zero. The integral over (c, ∞) is estimated by

$$\left| \int_c^\infty (k(s, t), n_j^+)_N \exp(-\mu_j s)\, ds \right| \le |f_+|_{H_0} |n_j^+|_N \frac{\exp(-\mu_j c)}{(2\mu_j)^{1/2}}$$

and can be made arbitrarily small by choosing c sufficiently large. It follows that

$$\lim_{t \to \infty} (g_1''(t), f_j^+)_E = 0 \qquad \text{for } j = 1, \cdots, m.$$

Since P is finite dimensional and spanned by the $\{f_j^\pm\}$ we see that

$$(4.9) \qquad\qquad \lim_{t \to \infty} g_1''(t) = 0.$$

Finally, we note that $\exp(-iA^2 t)$ is unitary on H' and bounded in norm on H; therefore combining (4.8) and (4.9) we get

(4.10) $W_+{}^S f_+ = \lim_{t\to\infty} \exp\,(-iA^2 t) g_1{}'(t)$.

The proof from here on is exactly the same as in the proof of Theorem 4.2 of Chapter II.

The Schrödinger scattering operator is defined as usual:

(4.11) $S^S = (W_+{}^S)^{-1} W_-{}^S$.

It is clear from Lemma 4.1 that S^S is unitary on H_0 and that

(4.12) $S^S f_+ = S f_+$ and $S^S f_- = S^{-1} f_-$;

likewise

(4.13) $S^S \tilde{f}_+ = S \tilde{f}_+$ and $S^S \tilde{f}_- = S^{-1} \tilde{f}_-$.

This apparent discrepancy between S^S and S stems from the fact that we have expressed the Schrödinger scattering matrix as an operator on the spectral representation of A_0 rather than that of $A_0{}^2$. An easy way to verify this is to introduce the time reversal operator T:

$$T\{\,f_1\,,f_2\} = \{\,f_1\,,\,-f_2\}\,.$$

It is clear that T commutes with A^2 and $A_0{}^2$ so that

(4.14) $S^S T = T S^S$.

On the other hand the action of T on the free space spectral representation can be easily computed; in fact we have by Theorem 2.1 of Chapter IV

$$\tilde{f}_0(\sigma\,,\,\omega) = (2\pi)^{-3/2}(\,f\,,\,\phi_{\sigma,\omega})_{H_0}$$

where

$$\phi_{\sigma,\omega}(x) = \sqrt{2}\,\exp\,(-i\sigma x\omega)\{1\,,\,i\sigma\}$$

and ω is a unit vector; in these terms

(4.15) $\overline{[Tf]}_0(\sigma\,,\,\omega) = \tilde{f}_0(-\sigma\,,\,-\omega)$.

The relations (4.14) and (4.15) show that S^S is even in σ, ω and hence its action is determined by only its values on the positive real axis.

Now the spectral representation for the operator $A_0{}^2$ has its support on the positive real axis and if we denote the energy parameter by τ, then τ

is related to the A_0 parameter σ by:

(4.16) $\tau = \sigma^2$.

According to (4.15) data of the form $\{0, f_2\}$ will have an odd spectral representation and hence, taking (4.16) into account, its spectral representer $\tilde{f}_S(\tau)$ in the A_0^2 representation is

$$\tilde{f}_S(\tau) = \tau^{-1/4}\tilde{f}_0(\sqrt{\tau})   (\tau > 0) ;$$

here the factor $\tau^{-1/4}$ has been introduced to preserve the isometric character of the mapping. In this representation the Schrödinger scattering matrix is simply

$$\mathbb{S}^S(\tau) = \mathbb{S}(\sqrt{\tau}) ,$$

restricted of course to the spectrum which is $[0, \infty)$.

Combining the above with the results of the preceding section we obtain:

Theorem 4.1.   *The Schrödinger scattering matrix $\mathbb{S}^S(\tau)$ has an analytic extension which is holomorphic in the 'physical plane,' except for poles at the bound state energies, and meromorphic in the 'nonphysical' plane.*

5. Notes and Remarks

Part I contains an initial attempt to apply our theory to symmetric hyperbolic systems. Our requirement that

$$a(\xi) = \sum a^i \xi_i$$

be invertible for all $\xi \neq 0$ is, we feel, too restrictive since Maxwell's equations, which are amenable to our theory (see Appendix 4), are not subsummed under this condition.

In part II we have assumed that zero is not an eigenvalue of the generator A for the acoustic equation with a potential; this was done only for expository reasons since the theory can also be adapted to handle this problem when zero is an eigenvalue of A (see Lax and Phillips [6]). Working independently and concurrently Dolph et al. [1] have proved for a wider class of potentials q, of growth $O(\exp(-\alpha |x|))$ for large $|x|$, but without boundary, that the scattering matrix is meromorphic for $|\operatorname{Im} z| < \alpha/2$ with poles occurring in the lower half-plane at most at the points $\{-i\mu_j\}$; the bound state energies for the corresponding Schrödinger operator are the numbers $\{-\mu_j^2\}$.

APPENDIX 1

Semigroups of Operators

We now establish those results from the theory of semigroups of operators which have been used in our development of scattering theory. In order to simplify the presentation we shall limit ourselves to semigroups of contraction operators on a Hilbert space H. Thus we will assume that $\{Z(t)\,;\, t \geq 0\}$ is a one-parameter family of operators on H satisfying the following properties:

(a) $Z(0) = I$ and $Z(t_1 + t_2) = Z(t_1)Z(t_2)$ for t_1, $t_2 > 0$;
(b) $|Z(t)| \leq 1$ for all $t > 0$;
(c) $\lim_{t \to 0+} Z(t)x = x$ for all x in H.

Lemma 1. $Z(t)x$ is continuous in $t \geq 0$ for all x in H.

Proof. By property (a) we can write

$$Z(t_2)x - Z(t_1)x = Z(t_1)[Z(t_2 - t_1)x - x] \qquad \text{for} \quad 0 \leq t_1 \leq t_0 \leq t_2\,;$$

and this converges to 0 as t_1, $t_2 \to t_0$ by properties (b) and (c).

The *infinitesimal generator* B is defined as

$$(1) \qquad Bx = \lim_{\Delta \to 0+} \frac{Z(\Delta)x - x}{\Delta}$$

and the domain of B consists of all those vectors for which this limit exists. It is clear that B is a linear operator on its domain $D(B)$.

Lemma 2. For any x in $D(B)$ the derivative of $Z(t)x$ exists strongly and

$$(2) \qquad \frac{dZ(t)x}{dt} = Z(t)Bx = BZ(t)x \qquad \text{for all} \quad t > 0\,.$$

245

Proof. If $\Delta > 0$, then by property (a)

$$\frac{Z(t + \Delta)x - Z(t)x}{\Delta} = Z(t)\frac{Z(\Delta)x - x}{\Delta} = \frac{Z(\Delta) - I}{\Delta}Z(t)x .$$

When x belongs to $D(B)$ the middle member of the above equality converges to $Z(t)Bx$ and hence the first and third members also converge; the first converging to the right derivative of $Z(t)x$ and the third to $BZ(t)x$. This shows incidently that $Z(t)D(B) \subset D(B)$. On the other hand if $\Delta > 0$ and $t - \Delta > 0$ then

$$\frac{Z(t)x - Z(t - \Delta)x}{\Delta} = Z(t - \Delta)\frac{Z(\Delta)x - x}{\Delta} .$$

Since $Z(t - \Delta)$ converges strongly to $Z(t)$, the right member of this equality converges to $Z(t)Bx$ while the left member converges to the left derivative of $Z(t)x$. It follows that the derivative of $Z(t)x$ exists and satisfies the relation (2).

Integrating $dZ(t)x/dt$ over the interval $[0, t]$ we get

Corollary 1. *If x belongs to $D(B)$ then*

$$(3) \qquad\qquad Z(t)x - x = \int_0^t Z(\tau)Bx \, d\tau .$$

More generally we have

Lemma 3. *For any x in H*

$$(4) \qquad\qquad Z(t)x - x = B\int_0^t Z(\tau)x \, d\tau .$$

Proof. Making use of property (a) we get

$$\frac{Z(\Delta) - I}{\Delta}\int_0^t Z(\tau)x \, d\tau = \frac{1}{\Delta}\int_0^t [Z(\tau + \Delta)x - Z(\tau)x] \, d\tau$$

$$= \frac{1}{\Delta}\int_t^{t+\Delta} Z(\tau)x \, d\tau - \frac{1}{\Delta}\int_0^\Delta Z(\tau)x \, d\tau ,$$

from which (4) follows.

Lemma 4. *The generator B is a closed linear operator with dense domain.*

Proof. According to Lemma 3

$$x_\Delta \equiv \frac{1}{\Delta} \int_0^\Delta Z(\tau) x \, d\tau$$

belongs to $D(B)$ for arbitrary x and since $\lim_{\Delta \to 0} x_\Delta = x$ we conclude that $D(B)$ is dense in H. On the other hand suppose that the sequence $\{x_n\}$ belongs to $D(B)$, $x_n \to x$ and $Bx_n \to y$. Then $Z(\tau)Bx_n$ converges to $Z(\tau)y$ uniformly in τ and it follows from the relation (3) that

$$\frac{Z(\Delta)x - x}{\Delta} = \frac{1}{\Delta} \int_0^\Delta Z(\tau) y \, d\tau .$$

Passing to the limit as Δ tends to 0 we see that x lies in $D(B)$ and that $Bx = y$; this proves that B is a closed operator.

Lemma 5. *A semigroup of operators is uniquely determined by its infinitesimal generator.*

Proof. Suppose the semigroups $\{Z_1(t)\}$ and $\{Z_2(t)\}$ have the same infinitesimal generator B. Then for x in $D(B)$, $Z_2(\tau)x$ belongs to $D(B)$ and therefore

$$\frac{d}{d\tau} Z_1(t - \tau) Z_2(\tau) x = Z_1(t - \tau)[BZ_2(\tau)x] + Z_1(t - \tau)[-B]Z_2(\tau)x = 0 .$$

Integrating this over the interval $[0, t]$ we get

$$Z_2(t)x - Z_1(t)x = 0 \qquad \text{for all} \quad t > 0 ;$$

and since $D(B)$ is dense in H we see that the two semigroups are identical.

The spectrum of the generator of a semigroup of contraction operators is always contained in the left half-plane; in fact, as the next lemma shows, for all λ with $\mathrm{Re}\,\lambda > 0$ the resolvent exists and can be expressed as the Laplace transform of the semigroup. We mean by the *resolvent* the operator

$$R(\lambda, B) = (\lambda I - B)^{-1} .$$

Lemma 6. *If* $\mathrm{Re}\,\lambda > 0$ *then*

(5)
$$R(\lambda, B)x = \int_0^\infty e^{-\lambda t} Z(t) x \, dt .$$

Proof. If in Lemma 3 and Corollary 1 we replace $\{Z(t)\}$ by the semi-group $\{\exp(-\lambda t)Z(t)\}$ (and hence B by $(B - \lambda I)$), then these results can be rewritten as

$$x - e^{-\lambda T}Z(T)x = (\lambda I - B)\int_0^T e^{-\lambda t}Z(t)x\,dt \qquad \text{for all} \quad x\,;$$

$$x - e^{-\lambda T}Z(T)x = \int_0^T e^{-\lambda t}Z(t)(\lambda I - B)x\,dt \qquad \text{for all} \quad x \quad \text{in} \quad D(B)\,.$$

It is clear that the integral

$$R(\lambda)y \equiv \int_0^\infty e^{-\lambda t}Z(t)y\,dt$$

converges to a bounded operator when $\operatorname{Re}\lambda > 0$. Hence passing to the limit as T becomes infinite and making use of the fact that B is a closed operator, we obtain

$$x = (\lambda I - B)R(\lambda)x \qquad \text{for all} \quad x\,;$$

$$x = R(\lambda)(\lambda I - B)x \qquad \text{for all} \quad x \quad \text{in} \quad D(B)\,.$$

This proves that λ is in the resolvent set for B and that $R(\lambda) = R(\lambda, B)$.

The resolvent of B satisfies the *resolvent equation:*

(6) $$R(\lambda, B) - R(\mu, B) = (\mu - \lambda)R(\lambda, B)R(\mu, B)$$

for all λ, μ in the resolvent set. In order to verify this it suffices to multiply

$$(\mu - \lambda)I = (\mu I - B) - (\lambda I - B)$$

on the right by $R(\mu, B)$ and on the left by $R(\lambda, B)$. It follows from (6) that the resolvents of B commute.

Lemma 7. *The resolvent set is open and the resolvent $R(\lambda, B)$ is a holomorphic function of λ on the resolvent set.*

Proof. If λ_0 belongs to the resolvent set of B then

$$(\lambda I - B)R(\lambda_0, B) = I + (\lambda - \lambda_0)R(\lambda_0, B)$$

and for

(7) $$|\lambda - \lambda_0| < |R(\lambda_0, B)|^{-1}$$

the right member is regular: hence the expression

$$R(\lambda_0, B)[I + (\lambda - \lambda_0)R(\lambda_0, b)]^{-1} = \sum_{n=0}^{\infty} (\lambda_0 - \lambda)^n [R(\lambda_0, B)]^{n+1}$$

is a right inverse for $(\lambda I - B)$ and a similar argument shows that it is also a left inverse. Thus for all λ sufficiently close to λ_0

$$(8) \qquad R(\lambda, B) = \sum_{n=0}^{\infty} (\lambda_0 - \lambda)^n [R(\lambda_0, B)]^{n+1}.$$

This explicit power series expansion for the resolvent proves that the resolvent set is open and that $R(\lambda, B)$ is holomorphic on the resolvent set.

The problem which occurs in most applications of this theory is to determine whether or not a given operator is the generator of a semigroup of operators. E. Hille and K. Yosida obtained the following solution to this problem:

Theorem 1. *A closed linear operator B with dense domain generates a strongly continuous semigroup of contraction operators if and only if*

$$(9) \qquad |R(\lambda, B)| \leq 1/\lambda \qquad \text{for all} \quad \lambda > 0.$$

Proof. The necessity follows directly from the expression (5):

$$|R(\lambda, B)x| \leq \int_0^{\infty} e^{-\lambda t} |Z(t)x| \, dt \leq |x| \int_0^{\infty} e^{-\lambda t} \, dt = |x|/\lambda$$

for all $\lambda > 0$.

In proving the converse we follow Yosida's argument and construct an approximating semigroup of the form

$$(10) \qquad Z_\lambda(t) \equiv \exp(tB_\lambda)$$

where

$$B_\lambda = \lambda^2 R(\lambda, B) - \lambda I.$$

To begin with we note that the inequality (9) implies

$$|\lambda R(\lambda, B)x - x| = |R(\lambda, B)Bx| \leq \lambda^{-1} |Bx|$$

if x belongs to $D(B)$. Thus for such x

$$(11) \qquad \lim_{\lambda \to \infty} \lambda R(\lambda, B)x = x$$

and since $D(B)$ is dense in H and the approximating operators $\{\lambda R(\lambda, B)\}$ are uniformly bounded by (9), the limit (11) exists for all x in H. In particular (11) shows that

$$B_\lambda x = \lambda^2 R(\lambda, B)x - \lambda(\lambda I - B)R(\lambda, B)x = \lambda R(\lambda, B)Bx$$

tends to Bx as λ becomes infinite for all x in $D(B)$.

Next we obtain a bound on the $\{Z_\lambda(t)\}$:

$$Z_\lambda(t) = \exp(-\lambda t) \exp(t\lambda^2 R(\lambda, B)) = \exp(-\lambda t) \sum_{n=0}^{\infty} \frac{(t\lambda^2)^n}{n!} [R(\lambda, B)]^n ;$$

and again making use of the inequality (9) we obtain

(12)
$$|Z_\lambda(t)| \le e^{-\lambda t} \sum_{n=0}^{\infty} \frac{(t\lambda)^n}{n!} = 1$$

for all $\lambda, t > 0$.

It is clear that all of the operators above commute. Hence

$$Z_\lambda(t) - Z_\mu(t) = \int_0^t \frac{d}{d\tau} [Z_\lambda(\tau) Z_\mu(t - \tau)] d\tau$$

$$= \int_0^t Z_\lambda(\tau) Z_\mu(t - \tau) (B_\lambda - B_\mu) d\tau .$$

Making use of the bound (12) we get

$$|Z_\lambda(t)x - Z_\mu(t)x| \le t |B_\lambda x - B_\mu x| .$$

This shows that

(13)
$$\lim_{\lambda \to \infty} Z_\lambda(t)x \equiv Z(t)x \qquad (t > 0) , ,$$

exists for each x in $D(B)$, uniformly on compact subsets of $[0, \infty)$. Since the $\{Z_\lambda(t)\}$ are uniformly bounded this limit exists for all x in H, again uniformly on compact subsets of $[0, \infty)$. Such a limit inherits from the approximating semigroups the properties (a), (b), and (c) and therefore $\{Z(t)\}$ defined by (13) is a strongly continuous semigroup of contraction operators.

Let C denote the generator of $\{Z(t)\}$. It remains to prove that $C = B$.

Now for x in $D(B)$ we have by Corollary 1

$$Z_\lambda(\Delta)x - x = \int_0^\Delta Z_\lambda(\tau)B_\lambda x \, d\tau \, .$$

Dividing by Δ and passing to the limit as λ tends to infinity we obtain

$$\frac{Z(\Delta)x - x}{\Delta} = \frac{1}{\Delta}\int_0^\Delta Z(\tau)Bx \, d\tau \, ,$$

from which it follows that $Cx = Bx$; that is $C \supset B$. For fixed $\lambda > 0$ the operator $(\lambda I - B)$ is by hypothesis one-to-one and onto and obviously no proper extension of $(\lambda I - B)$ can have these properties. On the other hand, according to Lemma 6 the operator $(\lambda I - C)$ does have these properties; consequently $C = B$ as desired.

The previous development is valid as proved for any Banach space. However, the following material on the adjoint semigroup holds as stated only for reflexive Banach spaces and we achieve some simplification in the argument by limiting ourselves to a Hilbert space.

For a semigroup of operators $\{Z(t)\}$ satisfying the properties (a), (b), and (c) we define the *adjoint semigroup* to be the family of adjoint operators:
$$\{Z^*(t) \; ; t \geq 0\} \, .$$

It is clear from the definition that the adjoint operators satisfy properties (a) and (b) and that

$$(Z^*(t)x \, , y) = (x \, , Z(t)y) \to (x \, , y)$$

as t tends to zero. Consequently,

$$|\, Z^*(t)x - x\,|^2 = (Z^*(t)x \, , Z^*(t)x) + (x \, , x) - (Z^*(t)x \, , x) - (x \, , Z^*(t)x)$$

has a nonpositive limit superior; this proves that the adjoint semigroup also satisfies property (c).

Lemma 8.   *The generator of the adjoint semigroup is the adjoint of the generator of the original semigroup.*

Proof.   Let C denote the generator of the adjoint semigroup and choose x and y in $D(B)$ and $D(C)$, respectively. Then

$$(Bx, y) = \lim_{\Delta \to 0}\left(\frac{Z(\Delta)x - x}{\Delta} \, , y\right) = \lim_{\Delta \to 0}\left(x, \frac{Z^*(\Delta)y - y}{\Delta}\right) = (x, Cy)$$

and therefore $C \subset B^*$. On the other hand if x and y belong to $D(B)$ and $D(B^*)$, respectively, then according to Corollary 1

$$(Z(\Delta)x - x, y) = \int_0^\Delta (BZ(\tau)x, y)\, d\tau = \int_0^\Delta (x, Z^*(\tau)B^*y)\, d\tau$$

and it follows from this that

$$Z^*(\Delta)y - y = \int_0^\Delta Z^*(\tau)B^*y\, d\tau .$$

Dividing by $\Delta > 0$ and passing to the limit as Δ tends to zero we see that

$$Cy = B^*y ,$$

which proves that $C \supset B^*$.

It is now easy to prove the Stone theorem characterizing the generators of groups of unitary operators. We recall that an operator A with dense domain is called *skew self-adjoint* if

$$A^* = -A .$$

Theorem 2.   *An operator generates a one-parameter strongly continuous group of unitary operators if and only if it is skew self-adjoint.*

Proof.   If $\{U(t)\}$ is a strongly continuous group of unitary operators with infinitesimal generator A then

$$\lim_{\Delta \to 0+} \frac{U(-\Delta)x - x}{\Delta} = \lim_{\Delta \to 0+} - U(-\Delta)\frac{U(\Delta)x - x}{\Delta}$$

exists if and only if x lies in $D(A)$ in which case the limit is simply $-A$. Since $U^*(t) = U(-t)$ this shows that the generator of the adjoint semigroup $\{U^*(t);\, t \geq 0\}$ is $-A$ and hence by Lemma 8 we have $A^* = -A$; in other words A is skew self-adjoint.

Conversely, if A is assumed to be skew self-adjoint then for any x in $D(A)$

$$(Ax, x) = (x, A^*x) = -(x, Ax)$$

so that (Ax, x) is pure imaginary-valued. Thus for

$$\lambda x \pm Ax = f \qquad (\lambda > 0) ,$$

the real part of the inner product with x is

$$\lambda(x,x) = \mathrm{Re}\,(x,f) \le |x|\,|f|\,;$$

this implies that $(\lambda I \pm A)$ is one-to-one and that

(14) $$|(\lambda I \pm A)^{-1}| \le 1/\lambda \quad \text{for} \quad \lambda > 0.$$

Since A is equal to the adjoint of $-A$ it is obviously closed and hence the range of $(\lambda I \pm A)$ is a closed subspace of H. Actually, the range is all of H for otherwise there would exist a nonzero y orthogonal to the range; that is

$$(\lambda x \pm Ax, y) = 0$$

for all x in $D(A)$. It follows from this that y belongs to $D(A^*) = D(A)$ and that

$$A^*y = -Ay = \mp\lambda y\,;$$

which is impossible because the only way $(Ay, y) = \pm\lambda(y, y)$ can then be pure imaginary is for y to be the zero vector. Thus $(\lambda I \pm A)$ is one-to-one and onto with $|(\lambda I \pm A)^{-1}| \le \lambda^{-1}$.

The Hille–Yosida theorem now asserts that A and $-A$ generate semigroups $\{U_+(t); t \ge 0\}$ and $\{U_-(t); t \ge 0\}$, respectively. Moreover, for x in $D(A)$

$$\frac{d}{dt}|U_\pm(t)x|^2 = (\pm AU_\pm(t)x, U_\pm(t)x) + (U_\pm(t)x, \pm AU_\pm(t)x) = 0$$

so that

$$|U_\pm(t)x| \equiv |x|\,;$$

the operators $\{U_\pm(t)\}$ are therefore isometries. Further for x in $D(A)$

$$\frac{d}{dt}U_+(t)U_-(t)x = U_+(t)AU_-(t)x + U_+(t)(-A)U_-(t)x = 0$$

so that

$$U_+(t)U_-(t) \equiv I.$$

Likewise,

$$U_-(t)U_+(t) \equiv I$$

and it follows that the operators $\{U_{\pm}(t)\}$ are unitary operators. Finally, it is clear from the approximating semigroups used in the proof of the Hille–Yosida theorem that all the operators in these two semigroups commute with one another; using this fact it is readily verified that

$$U(t) \; = \; \begin{cases} U_{+}(t) & \text{for} \quad t \geq 0 \\[2mm] U_{-}(-t) & \text{for} \quad t \leq 0 \end{cases}$$

defines a strongly continuous group of unitary operators with generator A.

APPENDIX 2

Energy Decay

The energy decay theorem plays a central role in Chapter V and there-fore it is of interest to explore alternative approaches to this result. We shall present two such alternatives in this appendix, the first of which is our original proof of the energy decay. Roughly speaking this argument goes as follows: If the energy does not decay then part of it will be trapped and this part behaves like a solution of the wave equation in a bounded domain in which case there are periodic solutions, contrary to Rellich's uniqueness theorem. In the second approach we prove directly that $\{U(t)\}$ has an absolutely continuous spectrum. This obviates the use of Theorem 2.2 and Lemma 2.3 in the energy decay proof given in Chapter V; in addi-tion, it provides the basis of still another proof of Rellich's uniqueness theorem.

Theorem 1. *If G' is a bounded subdomain of G, then*

$$(1) \qquad\qquad \liminf_{t\to\infty} |\, U(t)f\,|_{E^{G'}} = 0$$

for all f in H.

It is enough to prove this assertion for a dense subset of H, so we may as well assume that f belongs to $D(A)$. Our proof centers about the follow-ing measure of decay:

$$(2) \qquad\qquad \gamma(f) \equiv \sup_{G'} \{\liminf_{t\to\infty} |\, U(t)f\,|_{E^{G'}}\}\,,$$

where the supremum is taken over all bounded subdomains G' of G. Even-tually we shall prove that $\gamma(f) = 0$; however, we begin by proving

Lemma 1. *If f is a nonzero data belonging to $D(A)$, then $\gamma(f) < |f|_E$.*

Proof of Lemma 1. Suppose that $\gamma(f) = |f|_E$. Let $G(k) = G \cap \{|x| < k\}$ and choose $\epsilon > 0$. Then it follows from our assumption that there exist constants $n(\epsilon)$ and $t(\epsilon)$ such that

$$| \, |U(t)f|_{E^{G(k)}} - |f|_E \, | < \epsilon$$

for all $k \geq n(\epsilon)$ and $t \geq t(\epsilon)$. Since $|U(t)f|_E = |f|_E$ this amounts to the assertion

$$(3) \qquad\qquad \lim_{n,\, t \to \infty} \, |U(t)f|_{E^{CG(k)}} = 0 \, ,$$

where $CG(k)$ denotes the complement of $G(k)$. On the other hand

$$|U(t)f|_E + |AU(t)f|_E = |f|_E + |Af|_E$$

for all t and hence it follows from Theorem 1.4 of Chapter V that the set $\{U(t)f;\ -\infty < t < \infty\}$ is precompact in the local energy norm $|\cdot|_{E^{G(k)}}$. Thus given a sequence $\{t_n\}$ tending to infinity, we can choose a subsequence (which we renumber) by the diagonal process such that $\{U(t_n)f\}$ is a Cauchy sequence in each of the local energy norms $|\cdot|_{E^{G(k)}}$. Combining this with (3) we see that $\{U(t_n)f\}$ is also a Cauchy sequence in the H norm. This can be used to prove the analogous assertion for a sequence $\{t_n\}$ tending to minus infinity since

$$(U(t_k)f,\, U(t_n)f)_E = (U(-t_n)f,\, U(-t_k)f)_E \, ,$$

from which it follows that

$$|U(t_k)f - U(t_n)f|_E = |U(-t_k)f - U(-t_n)f|_E \, .$$

Finally, if $\{t_n\}$ has a finite limit then the strong continuity of the group of operators implies that $\{U(t_n)f\}$ is a Cauchy sequence.

The above establishes the fact that $U(t)f$ is an *almost periodic* vector-valued function and as such is the superposition of a denumerable set of exponentials $a_j \exp(i\sigma_j t)$, and

$$a_j = \lim_{T \to \infty} \frac{1}{2T} \int_{-T}^{T} \exp(-i\sigma_j t)\, U(t)f \, dt$$

in the H topology. Since A is a closed operator and the functions $U(t)f$

and $AU(t)f = U(t)Af$ are strongly continuous, we may write

$$A \int_{-T}^{T} e^{-i\sigma t} U(t) f \, dt = \int_{-T}^{T} e^{-i\sigma t} A U(t) f \, dt = \int_{-T}^{T} e^{-i\sigma t} \frac{dU(t)f}{dt} \, dt$$

$$= e^{-i\sigma T} U(T) f - e^{i\sigma T} U(-T) f + i\sigma \int_{-T}^{T} e^{-i\sigma t} U(t) f \, dt .$$

Dividing by $2T$ and passing to the limit as T becomes infinite we see that a_j belongs to $D(A)$ and that $A a_j = i\sigma_j a_j$. This shows that σ_j belongs to the point spectrum of A which was proved to be empty in Theorem 2.2 of Chapter V. We conclude that $f = 0$.

Proof of Theorem 1. Suppose now that $\gamma(f) > 0$ for some f in $D(A)$. By (2) there exist sequences $\{t_n\}$ and $\{k_n\}$, both tending to infinity, such that

$$|\ |U(t_n) f|_{E^{G(k_n)}} - \gamma(f)\ | < 1/n .$$

As in the proof of the above lemma, we make use of the local compactness of the sequence $\{U(t_n)f\}$ to abstract a subsequence $\{t_n'\}$ such that $\{U(t_n')f\}$ converges in the local energy norms to, say, ϕ. It is clear that $|\phi|_E = \gamma(f)$. In order to verify that ϕ belongs to H we note that $\{U(t_n')f\}$, being bounded in norm, contains a subsequence which converges weakly to some vector ψ in H which can be identified with ϕ.

For any $s > 0$ and $l \geq s$, a domain of dependence argument (Theorem 1.1 of Chapter V) shows that for all p

$$|\ U(t_n' + s)f - U(s)\phi\ |_{E^{G(p)}} \leq |\ U(t_n')f - \phi\ |_{E^{G(p+l)}} .$$

Consequently,

$$|\ U(s)\phi\ |_{E^{G(p)}} \geq |\ U(t_n' + s)f\ |_{E^{G(p)}} - |\ U(t_n')f - \phi\ |_{E^{G(p+l)}}$$

and passing to the limit as n becomes infinite we see that

$$|\ U(s)\phi\ |_{E^{G(p)}} \geq \liminf_{t \to \infty} |\ U(s)f\ |_{E^{G(p)}} .$$

It follows that $\gamma(\phi) \geq \gamma(f)$. However, $\gamma(\phi) \leq |\phi|_E$ and $\gamma(f) = |\phi|_E$ so that $\gamma(\phi) = |\phi|_E$; this contradicts the lemma and thus concludes the proof of Theorem 1.

The second proof of the energy decay is based on the following result:

Theorem 2. *The spectrum of A is absolutely continuous.*

Proof. As we have already indicated in Remark 2.2 of Chapter V, the wave operators $W_\pm$ exist and are intertwining operators for the two groups of unitary operators; that is

(4) $$U(t)W_\pm = W_\pm U_0(t) \ .$$

It follows from this that $U(t)$ restricted to the range of W_+ [or W_-] is unitarily equivalent with $U_0(t)$. Since by Theorem 2.1 of Chapter IV, $\{U_0(t)\}$ has an absolutely continuous spectrum, so does $\{U(t)\}$ on the closure R of the sum of the ranges of W_+ and W_-. Further, W_+ [W_-] restricted to $D_+{}^\rho$ [$D_-{}^\rho$] acts like the identity and consequently R contains $D_+{}^\rho + D_-{}^\rho$.

Assertion. $R = H$. Clearly, Theorem 2 follows from this assertion; to prove it assume on the contrary that there exists a nonzero vector f orthogonal to R. Since the ranges of W_+ and W_- are invariant under $\{U(t)\}$ so is R and the orthogonal complement to R. Consequently, $U(t)f$ remains orthogonal to R and hence to $D_+{}^\rho + D_-{}^\rho$ for all t. We deduce from this by means of the next lemma that $[U(t)f](x)$ will be zero for $|x| > \rho$ and all t, and by employing the Holmgren uniqueness theorem (Theorem 1.5 of Chapter IV) we conclude that $[U(t)f](x)$ vanishes in all of G.

Lemma 2. *If $U(t)f$ is orthogonal to $D_+{}^\rho + D_-{}^\rho$ for all t, then $[U(t)f](x)$ equals zero for all $|x| > \rho$ and all t.*

Proof. For each real t and $r > \rho$ let $u_m(r, t)$ denote the mth spherical harmonic coefficient of $[U(t)f]_1$. According to Theorem 3.3 of Chapter IV, each such function defines an analytic element in the strip $|\operatorname{Re} t - t_0| < \delta$ for each $r > \rho + \delta$. Furthermore, if $|t_1 - t_2| < \delta$, t_1 and t_2 real, then the two analytic elements corresponding to t_1 and t_2 must coincide for each real t between t_1 and t_2 and for almost all $r > \rho + \delta$; actually they will coincide for all $r > \rho + \delta$ since the elements are also analytic in r for $r > \rho$. Thus, because of their analyticity the two elements coincide for each $r > \rho + \delta$ in the common strip $t_1 < \operatorname{Re} t < t_2$ (assuming that $t_1 < t_2$). It follows that the above elements define an entire function in t for each $r > \rho + \delta$. A glance at the expression (3.15) in Chapter IV shows that

$u_m(r, t)$ is of the order of $|t|^{m+n-3}$ in the cut plane and since $u_m(r, t)$ is entire in t it must really be a polynomial in t of degree less than $m + n - 2$.

We now make use of the fact that the energy of $U(t)f$ is uniformly bounded in the variable t. In particular the energy in each coefficient in the region $r > \rho + \delta$ must be bounded and at the same time of polynomial growth. This requires $[U(t)f](x)$ to be constant in time for $|x| > \rho + \delta$; that is $[U(t)f](x) = \{g_1(x), g_2(x)\}$ where $g_2(x) = 0$ and $g_1(x)$ is harmonic. Since g_1 also has a finite Dirichlet integral it must be equal to some constant c for $|x| > \rho + \delta$. Applying the inequality (1.2) of Chapter IV with $f_1 = g_1$ we obtain

$$c^2 r^{n-2} \leq \text{const} \, |f|_E^2$$

for all $r > \rho + \delta$, from which it follows that $g_1(x) = 0$ for $|x| > \rho + \delta$. Since $\delta > 0$ was arbitrary, this completes the proof of Lemma 2.

It now follows directly from the integral representation theorem for $\{U(t)\}$ in terms of its resolution of the identity and the Riemann–Lebesgue theorem that $\lim_{t \to \infty} (U(t)f, g)_E = 0$ for all f, g in H. This extends Lemma 2.3 of Chapter V.

Remark. The only difficult part of our proof of the Rellich uniqueness theorem involves showing that A has no point spectrum. Consequently, the above theorem combined with the proof of Theorem 2.3 of Chapter V, gives another proof of the Rellich theorem.

APPENDIX 3

Energy Decay for Star-Shaped Obstacles

Cathleen S. Morawetz *

The standard energy conservation law for U is found by multiplying the wave equation by U_T and noting that the resulting quadratic expression is a divergence:

$$(1) \qquad U_T(U_{TT} - \Delta_X U) = \operatorname{div}_X P + Q_T$$

where

$$(2) \qquad P = -U_T \nabla U , \qquad Q = \tfrac{1}{2}(U_T^2 + |\nabla U|^2) .$$

Integrated over any region $\mathfrak{D}$ this expression therefore yields a surface integral in (X, T) space which vanishes whenever U is a solution of the wave equation; this is called the standard energy identity. It has the additional property that the integrand is a positive definite form on spacelike surfaces.

As is well known, the Kelvin transformation

$$(3) \quad X = \frac{x}{r^2 - t^2} , \quad T = \frac{t}{r^2 - t^2} , \quad RU = ru , \quad R = |X| , \quad r = |x| ,$$

preserves the wave operator in the sense that

$$(4) \qquad R^3(U_{TT} - \Delta_X U) = r^3(u_{tt} - \Delta_x u) .$$

On the other hand,

$$(5) \qquad R U_T = r[(r^2 + t^2)u_t + 2t(ru)_r]$$

* Department of Mathematics, New York University, New York.

261

and

$$(6) \qquad \frac{dX \, dT}{R^4} = \frac{dx \, dt}{r^4}$$

Combining (4), (5), and (6) we get

$$(7) \qquad \int U_T(U_{TT} - \Delta_X U) \, dX \, dT = \int Nu(u_{tt} - \Delta_x u) \, dx \, dt$$

with $Nu = (r^2 + t^2)u_t + 2t(ru)_r$.

Using (1), the left hand side of (7) can be written as a surface integral and therefore so can the right side. Thus one obtains:

Theorem 1. *Suppose* $u(x, t)$ *is a solution of the wave equation which has square integrable derivatives. Then over any three dimensional surface* ∂ *with the surface element* dS,

$$(8) \qquad \int_{\partial} (pn + qn_t) \, dS = 0$$

where n *is the space component of the outward normal,* n_t *is the time component.*

A tedious calculation gives

$$(9) \qquad p = -tu_t^2 x - 2t(x \, \nabla u) \, \nabla u + t \mid \nabla u \mid^2 x - (r^2 + t^2)u_t \, \nabla u - 2tu \, \nabla u$$

$$- \tfrac{1}{2}r^{-2}((r^2 + t^2)u^2)_t \, x$$

$$(10) \qquad q = 2t(x\nabla u)u_t + \tfrac{1}{2}(r^2 + t^2)(\mid \nabla u \mid^2 + u_t^2) + 2tuu_t$$

$$+ r^{-2}(r^2 + t^2)(u \, \nabla u \, x + \tfrac{1}{2}u^2).$$

q *is a definite form;* in fact, q may be written as:

$$(11) \qquad q = \tfrac{1}{2}(r^2 + t^2)(\mid \nabla u \mid^2 - u_r^2)$$

$$+ \frac{1}{4r^2} \{(r + t)^2((ru)_r + (ru)_t)^2 + (r - t)^2((ru)_r - (ru)_t)^2\}.$$

The positivity of q can also be deduced as follows: Under the Kelvin transformation, the inverse of (3), the surface $t = $ constant is transformed into a spacelike surface in the (X, T) space. On this spacelike surface the integrand in the standard energy identity is a definite form. Hence on the transform of this surface, i.e., $t = $ constant, the new integrand, q, is also definite.

Theorem 2. *Let u be a solution of the wave equation outside a star-shaped body with boundary B and assume that $u = 0$ on B. Suppose further that the initial data f of u vanish for $|x| \geq k$. Then*

$$(12) \qquad |u(\tau)|_h \leq \frac{2k}{\tau}|f|$$

for $\tau \geq 2h$. Here $|u(\tau)|_h^2$ is the energy, $\int (|\nabla u| + u_t^2)\,dx$, inside a sphere of radius h at time τ and $|f|^2$ is the total energy of the initial data.

In the notation of Chapter V, $|f|_E = |f|$, $|U(\tau)f|_E^{G(h)} = |u(\tau)|_h$ so that the estimate (3.4) of Chapter V follows.

Proof. Choose the origin so that B is star-shaped with respect to the origin, i.e., $xn \leq 0$. We apply Theorem 1 to a domain bounded by the planes $t = \tau$, $t = 0$ and the body cylinder $x \in B$, $0 \leq t \leq \tau$. Then since the solution vanishes for r large enough,

$$(13) \qquad \int_{t=\tau} q\,dx + \int_0^\tau \int_B pn\,ds\,dt = \int_{t=0} q\,dx.$$

Since u vanishes on B, $\nabla u = (\partial u/\partial n)n$ there and $u_t = 0$; thus from (9), it follows that $pn = -t(\partial u/\partial n)^2 xn$. Since B is star-shaped with respect to the origin, $xn \leq 0$; thus $pn \geq 0$. Hence from (13),

$$(14) \qquad \int_{t=\tau} q\,dx \leq \int_{t=0} q\,dx.$$

From the expression (11) for q, we see that for $t = 0$, $q \leq \frac{1}{2}r^2(|\nabla u|^2 + u_t^2)$. Therefore, since f has support in $|x| \leq k$, we find

$$(15) \qquad \int_{t=0} q\,dx \leq \frac{1}{2}k^2|f|^2.$$

Since the integrand q is positive, we get from (14) and (15) for any h,

$$(16) \qquad \int_{\substack{t=\tau \\ r \leq h}} q\,dx \leq \int_{t=\tau} q\,dx \leq \frac{1}{2}k^2|f|^2.$$

Using the expression (11) we can bound q from below for $r \leq t/2$:

(17)

$$\tfrac{1}{4}t^2 \left[\tfrac{1}{2}(|\nabla u|^2 - u_r^2) + \frac{1}{4r^2}((ru)_r + (ru)_t)^2 + \frac{1}{4r^2}((ru)_r - (ru)_t)^2 \right] \leq q$$

or

(18)
$$\tfrac{1}{8}t^2 \left(|\nabla u|^2 + u_t^2 + \operatorname{div} \frac{1}{r^2} u^2 x \right) \leq q .$$

Using (18) in (16), we get for $\tau \geq 2h$

(20)
$$\int_{\substack{t=\tau \\ r \leq h}} \left(|\nabla u|^2 + u_t^2 + \operatorname{div} \frac{1}{r^2} u^2 x \right) dx \leq \frac{4}{\tau^2} k^2 |f|^2;$$

since $u = 0$ on B, integrating the divergence gives

(21)
$$\int_{\substack{t=\tau \\ r \leq h}} (|\nabla u|^2 + u_t^2) \, dx + \int_{\substack{t=\tau \\ r = h}} \frac{1}{r} u^2 \, dS \leq \frac{4}{\tau^2} k^2 |f|^2$$

where dS is the surface element on the sphere $r = h$. Hence,

(22)
$$\int_{\substack{t=\tau \\ r \leq h}} (|\nabla u|^2 + u_t^2) \, dx \leq \frac{4}{\tau^2} k^2 |f|^2$$

which concludes the proof.

APPENDIX 4

Scattering Theory for Maxwell's Equations

Georg Schmidt *

This appendix discusses the scattering of electromagnetic waves by an obstacle. The propagation of electromagnetic waves is governed by Maxwell's equations:

$$e_t = \operatorname{curl} m, \qquad m_t = - \operatorname{curl} e \; ;$$

these constitute a symmetric system of first order equations; yet this system does not come under the theory developed in Chapter VI since it has modes which propagate with zero speed. However, if we limit ourselves to divergence free fields, these modes are not excited and the general theory of scattering is applicable.

For free fields, i.e., propagation in whole space, the theory of the initial value problem for Maxwell's equations can be reduced to that for the wave equation by the introduction of a vector potential. Incoming and outgoing solutions are defined as before and the corresponding free translation representation can be constructed explicitly. For the perturbed problem, i.e., in the presence of an obstacle, we proceed as follows: Denote by B the operator

$$B = \begin{pmatrix} 0 & \operatorname{curl} \\ -\operatorname{curl} & 0 \end{pmatrix}$$

and abbreviate $\{e, m\}$ as f. Integration by parts gives

$$(f, Bf) = \int \det (n, e, m) \, dS,$$

* Mathematics Research Center, University of Wisconsin, Madison, Wisconsin.

n the normal to the boundary. This shows that energy is conserved for fields where the boundary values of e and m and the normal n are linearly dependent. There are two types of linear boundary conditions which will assure this:

(1) $\alpha e + \beta m$ is parallel to n, α and β given functions on the boundary.

(2) e and m are linearly dependent.

It is not hard to show that if we take the domain of the operator B to consist of smooth functions subject to one of the boundary conditions of type (1) or (2), the *closure* of the resulting operator, which we also denote by B, is skew self-adjoint. Since all states within the nullspace of B propagate with speed zero, we restrict our attention to the orthogonal complement of the nullspace of B.

The argument presented in Chapter VI shows that B has no point spectrum outside of zero. To prove energy decay we need to know the local compactness of the set of functions f which are orthogonal to the nullspace of B and which satisfy

$$|f| + |Bf| \leq 1 .$$

We shall deduce this in the special case where the boundary condition is of form (1), with α and β constant. In this case we have the following

Lemma.   *Suppose that $f = \{e, m\}$ belongs to the domain of B and is orthogonal to the nullspace of B. Then e and m are divergence free and satisfy the following boundary conditions:*

(a)   *$\alpha e + \beta m$ is parallel to n on the boundary.*

(b)   *$\beta e - \alpha m$ is orthogonal to n on the boundary.*

Proof.   To prove that e and m are divergence free we use the fact that if ϕ and ψ are arbitrary smooth functions with compact support,

$$f = \{\operatorname{grad} \phi , \operatorname{grad} \psi\}$$

belongs to the nullspace of B. Since ϕ and ψ vanish on the boundary integration by parts gives

$$(\{e , m\} , f) = (\operatorname{div} e , \phi) + (\operatorname{div} m , \psi) .$$

Since ϕ and ψ are arbitrary, it follows that $\operatorname{div} e = \operatorname{div} m = 0$.

The boundary condition (a) is imposed on the domain of B. To prove (b) we note that even when ϕ is not zero on the boundary, every f of the form

$$f = \{\beta \operatorname{grad} \phi, \; -\alpha \operatorname{grad} \phi\},$$

belongs to the domain of B and is annihilated by B. Integrating by parts and using $\operatorname{div} e = \operatorname{div} m = 0$ we get

$$(\{e, m\}, f) = \int (\beta e - \alpha m) \cdot n \, \phi \, dS.$$

Since the values of ϕ on the boundary are arbitrary it follows that every $\{e, m\}$ orthogonal to the nullspace of B satisfies the boundary condition (b).

Next we appeal to an inequality of Friedrichs [2], according to which

$$| \partial_x g | \le \operatorname{const} [| g | + | \operatorname{curl} g | + | \operatorname{div} g |]$$

for every vector-valued function g which is either parallel or perpendicular to n on the boundary. According to the lemma for $\{e, m\}$ orthogonal to the nullspace of B we can apply this inequality to $g = \alpha e + \beta m$ and to $g = \beta e - \alpha m$ separately. The resulting inequalities imply

$$| \partial_x e | + | \partial_x m | \le \operatorname{const} [| e | + | m | + | \operatorname{curl} e | + | \operatorname{curl} m |$$
$$+ | \operatorname{div} e | + | \operatorname{div} m |].$$

Since e and m are divergence free we can rewrite this for $f = \{e, m\}$ as

$$| \partial_x f | \le \operatorname{const} [| f | + | Bf |].$$

The desired local compactness now follows by the Rellich compactness criterion.

The rest of the theory proceeds as in Chapter VI and yields the meromorphic character in the whole complex plane of the scattering matrix for Maxwell's equations. The poles of the scattering matrix are those values of z for which the reduced equation $Bf = izf$ has a nontrivial outgoing solution satisfying the boundary conditions.

References

Adamjan, V. M., and Arov, D. Z.

[1] A class of scattering operators and of characteristic operator-functions of contractions, *Dokl. Akad. Nauk SSSR* **160,** 9–12 (1965); Engl. transl. in *Soviet Math. Dokl.* **6,** 1–5 (1965).

[2] Scattering operators and contraction semigroups in Hilbert space, *Dokl. Akad. Nauk SSSR* **165,** 9–12 (1965); Engl. transl. in *Soviet Math. Dokl.* **6,** 1377–1380 (1965).

Adams, F., Lax, P. D., and Phillips, R. S.

[1] On matrices whose real linear combinations are nonsingular, *Proc. Am. Math. Soc.* **16,** 318–322, (1965).

[2] Correction to "On matrices whose real linear combinations are nonsingular," *Proc. Am. Math. Soc.* **17,** 945–947 (1966).

Agmon, S.

[1] The coerciveness problem for integro-differential forms, *J. Analyse Math.* **6,** 183–223 (1958).

Aronszajn, N.

[1] On coercive integro-differential quadratic forms, Conference on partial differential equations, University of Kansas, Lawrence, Kansas, 1954, Tech. Rept. No. 14, pp. 94–106.

Beurling, A.

[1] On two problems concerning linear transformations in Hilbert space, *Acta Math.* **81,** 239–255 (1949).

Birman, M. Sh.

[1] Conditions for the existence of wave operators, *Dokl. Akad. Nauk SSSR* **143,** 506–509 (1962); Engl. transl. in *Soviet Math. Dokl.* **3,** 408–411 (1962).

[2] Existence conditions for wave operators, *Izv. Acad. Nauk SSSR, Ser. Mat.* **27,** 883–906 (1963).

Birman, M. Sh., and Entina, S. B.

[1] A stationary approach in the abstract theory of scattering, *Dokl. Akad. Nauk SSSR* **155,** 506–508 (1964); Engl. Transl. in *Soviet Math. Dokl.* **5,** 432–435 (1964).

Birman, M. Sh., and Kreĭn, M. G.

[1] On the theory of wave operators and scattering operators, *Dokl. Akad. Nauk SSSR* **144,** 475–478 (1962); Engl. transl. in *Soviet Math. Dokl.* **3,** 740–743 (1962).

Buchal, R. N.
    [1] The approach to steady state of solutions of exterior boundary value problems for the wave equation, *J. Math. Mech.* **12,** 225–234 (1963).

Cook, J. M.
    [1] Convergence of the Møller wave-matrix, *J. Math. Phys.* **36,** 82–87 (1957).

Courant, R., and Lax, P. D.
    [1] The propagation of discontinuities in wave motion, *Proc. Nat. Acad. Sci.* **42,** 872–876 (1956).

Dolph, C. L., McLeod, J. B., and Thoe, D.
    [1] The analytic continuation to the unphysical sheet of the resolvent kernel and the scattering operator associated with the Schrödinger operator, *J. Math. Anal. Appl.* **16,** 311–332 (1966).

Eidus, D. M.
    [1] On the principle of limiting absorption, *Mat. Sb., N. S.* **57,** (99), 13–44 (1962).

Faddeev, L. D.
    [1] Mathematical aspects of the three-body problem in the quantum scattering theory, *Tr. Mat. Inst. Steklov.* **69,** (1963) (Russian). Engl. transl. in. *Israel Program Sci. Translations,* Jerusalem, p. 110 (1965).

Foias, C., and Sz.-Nagy, B.
    [1] Contractions V. Translations bilatérales, *Acta Sci. Math.* **23,** 106–129 (1962).
    [2] Sur les contractions de l'espace de caractéristiques. Modèles fonctionneles, *Acta Sci. Math.* **25,** 38–71 (1964).

Fourès, Y., and Segal, I. E.
    [1] Causality and analyticity, *Trans. Am. Math. Soc.* **78,** 385–405 (1955).

Friedlander, F. G.
    [1] On the radiation field of pulse solutions of the wave equation II, *Proc. Royal Soc.,* **279A,** 386–394 (1964).

Friedrichs, K. O.
    [1] Ueber die Spektralzerlegung eines Integral-Operators, *Math. Ann.* **115,** 249–272 (1938).
    [2] Differential forms on Riemannian manifolds, *CPAM* **8,** 551–590 (1955).
    [3] Symmetric positive linear differential equations, *CPAM* **11,** 333–410 (1958).
    [4] "Perturbation of Spectra in Hilbert Space," in Lectures in Applied Mathematics, p. 178 Vol. 3. Am. Math. Soc., Providence, Rhode Island, 1965.

Gelfand, I. M., Graev, M. I., and Vilenkin, N. Ya.
    [1] "Generalized Functions, Vol. 5: Integral geometry and Representation Theory." Academic Press, New York, 1966.

Gelfand, I. M., and Levitan, B. M.
    [1] On the determination of a differential equation from its spectral function, *Izv. Akad. Nauk SSSR* **15,** 309–360 (1951).

Gell-Man, M., and Goldberger, M. L.
    [1] The formal theory of scattering, *Phys. Rev.* **91,** 398–408 (1953).

Halmos, P. R.
    [1] Shifts on Hilbert spaces, *Crell's J.* **208,** 102–112 (1961).

Heisenberg, W.
    [1] Die 'beobachtbaren Grössen' in der Theorie der Elementarteilchen, Part I, *Z. Physik* **120,** 513–538 (1943); Part II, **120,** 673–702 (1943).

Helgason, S.
[1] The Radon transform on Euclidean spaces, compact two-point homogeneous spaces and Grassmann manifolds, *Acta Math.* **113**, 153–180 (1965).
Helson, H.
[1] "Lectures on Invariant Subspaces," p. 130. Academic Press, New York, 1964.
Hille, E., and Phillips, R. S.
[1] "Functional Analysis and Semigroups" (rev. ed.), p. 808. Am. Math. Soc., Colloquium Pub., Vol. 31, 1957.
Hörmander, L.
[1] "Linear partial differential equations," p. 284. Springer, Berlin, 1963.
Ikebe, T.
[1] Eigenfunction expansions associated with the Schrödinger operator and their applications to scattering theory, *Arch. Rat. Mech. Analysis* **5**, 1–34 (1960).
Jauch, J. M.
[1] Theory of the scattering operator, *Helv. Phys. Acta* **31**, 127–158 (1958).
John, F.
[1] "Plane waves and spherical means applied to partial differential equations," p. 172. Wiley (Interscience), New York and London, 1955.
Kato, T.
[1] Perturbation of continuous spectra by trace class operators, *Proc. Japan Acad.* **33**, 260–264 (1957).
[2] Wave operators and unitary equivalence, *Pacific J. Math.* 171–180 (1965).
Kato, T., and Kuroda, S. T.
[1] A remark on the unitary property of the scattering operator, *Nuovo Cimento*, **14**, 1102–1107 (1959).
Kupradse, W. D.
[1] "Randwertaufgaben der Schwingungstheorie and Intergralgleichungen," p. 239. Akademie Verlag, Berlin, 1956.
Kuroda, S. T.
[1] On the existence and unitary property of the scattering operator, *Nuovo Cimento*, **12**, 431–454 (1959).
[2] An abstract stationary approach to perturbation of continuous spectra and scattering theory, to appear in *J. d'Analyse Math.*
Ladyzhenskaya, O. A.
[1] On the asymptotic amplitude principle, *Usp. Mat. Nauk* **12**, 161–164 (1957).
Lax, P. D.
[1] Asymptotic solutions of oscillatory initial value problems, *Duke Math. J.* **24**, 627–646 (1957).
[2] Translation invariant spaces, *Acta Math.* **101**, 163–178 (1959).
[3] Translation invariant spaces, *in* "Proc. Internat. Symp. Linear Spaces," pp. 299–306. Pergamon Press, Oxford, 1961.
Lax, P. D., Morawetz, C. S., and Phillips, R. S.
[1] Exponential decay of solutions of the wave equation in the exterior of a star-shaped obstacle, *CPAM* **16**, 477–486 (1963).
Lax. P. D., and Phillips, R. S.
[1] Local boundary conditions for dissipative symmetric linear differential operators, *CPAM* **13**, 427–455 (1960).
[2] The wave equation in exterior domains, *Bull. Am. Math. Soc.* **68**, 47–49 (1962).

[3] Scattering theory, *Bull. Am. Math. Soc.* **70**, 130–142 (1964).
[4] Analytic properties of the Schrödinger scattering matrix, *in* "Perturbation Theory and Its Applications in Quantum Mechanics" (Calvin H. Wilcox ed.), pp. 243–253. Wiley, 1966.
[5] Scattering theory for transport phenomena, to appear in Proc. Conf. on Functional Analysis, Univ. California at Irvine, California, 1966.
[6] The acoustic equation with an indefinite energy form and the Schrödinger equation, to appear in *J. Functional Analysis*.

Lippman, B. A., and Schwinger, J.
[1] Variational principles for scattering processes. I, *Phys. Rev.* **79**, 469–480 (1950).

Ludwig, D.
[1] Exact and asymptotic solutions of the Cauchy problem, *CPAM* **XIII**, 473–508 (1960).

Mackey, G. W.
[1] A theorem of Stone and von Neumann, *Duke Math. J.* **16**, 313–326 (1949).

Masani, P., and Robertson, J.
[1] The time-domain analysis of a continuous parameter weakly stationary stochastic process, *Pacific J. Math.* **12**, 1361–1378 (1962).

Mizohata, S.
[1] Sur l'Analyticité de la Fonction Spectrale de l'Opérateur Δ Relatif au Problème Extérieur, *Proc. Japan Acad.* **39**, 352–357 (1963).

Moeller, J. W.
[1] On the spectra of some translation invariant spaces, *J. Math. Analysis Applications* **4**, 276–296 (1962).

Møller, C.
[1] General properties of the characteristic matrix in the theory of elementary particles, *Kgl. Danske Vidensk. Selskab, Mat. fys. Medd.* **23**, 2–48 (1945).

Morawetz, C. S.
[1] The decay of solutions of the exterior initial-boundary value problem for the wave equation, *CPAM* **14**, 561–568 (1961).
[2] The limiting amplitude principle, *CPAM* **15**, 349–361 (1962).
[3] The limiting amplitude principle for arbitrary finite bodies, *CPAM* **18**, 183–189 (1965).

Müller, C.
[1] Randwertprobleme der Theorie elektromagnetischer Schwingungen, *Math. Z.* **56**, 261–270 (1952).
[2] "Grundprobleme der Mathematischen Theorie elektromagnetischer Schwingungen," p. 344, Springer, Berlin, 1957.

Phillips, R. S.
[1] Perturbation theory for semigroups of linear operators, *Trans. Am. Math. Soc.* **74**, 199–221 (1953).
[2] Dissipative operators and hyperbolic systems of partial differential equations, *Trans. Am. Math. Soc.* **90**, 193–254 (1959).

Povsner, A. Y.
[1] Expansion of arbitrary functions in terms of the eigenfunctions of the operator $- \Delta u + cu$, *Mat. Sbornik* **32** (74), 109–156 (1953).

Rellich, Fr.
[1] Ein Satz über mittlere Konvergenz, *Gött. Nachr. (math. phys.)*, 30–35 (1930).

[2] Über das asymptotische Verhalten der Lösungen von $\Delta u + \lambda u = 0$ in unendlichen Gebieten, *Jahresber. Deutsch. Math. Verein* **53**, 57–65 (1943).

Rosenblum, M.

[1] Perturbation of the continuous spectrum and unitary equivalence, *Pacific J. Math.* **7**, 997–1010 (1957).

Sinaĭ, Ja. G.

[1] Dynamical systems with countable Lebesgue spectrum. I, *Izv. Akad. Nauk SSSR* **25**, 899–924 (1961).

van Kampen, N. G.

[1] S-matrix and causality condition. I: Maxwell field, *Phys. Rev.* **89**, 1072–1079 (1953).

[2] S-matrix and causality condition. II: Nonrelativistic particles, *Phys. Rev.* **91**, 1267–1276 (1953).

von Schwarze, Günter,

[1] Über die 1., 2., und 3. äussere Randwertaufgabe der Schwingungsgleichung $\Delta F + k^2 F = 0$, *Math. Nachr.* **28**, 337–363 (1965).

Werner, P.

[1] Randwertprobleme der mathematischen Akustik, *Arch. Rat. Mech. Anal.* **10**, 29–62 (1962).

Weyl, H.

[1] Die natürlichen Randwertaufgaben im Aussenraum für Strahlungsfelder beliebiger Dimension und beliebigen Ranges, *Math. Z.* **56**, 105–119 (1952).

Yosida, K.

[1] "Functional Analysis," p. 458. Springer, Berlin, 1965.

Index

Numbers in italics indicate the pages on which the complete references are listed.